T0137267

Smart Cities

Schahram Dustdar · Stefan Nastić
Ognjen Šćekić

Smart Cities

The Internet of Things, People and Systems

Springer

Schahram Dustdar
TU Wien
Vienna, Austria

Ognjen Šćekić
TU Wien
Vienna, Austria

Stefan Nastić
TU Wien
Vienna, Austria

ISBN 978-3-319-86763-2   ISBN 978-3-319-60030-7   (eBook)
DOI 10.1007/978-3-319-60030-7

Printed on acid-free paper

This Springer imprint is published by Springer Nature
The registered company is Springer International Publishing AG
The registered company address is: Gewerbestrasse 11, 6330 Cham, Switzerland

"The best way to plan for downtown is to see how people use it today; to look for its strengths and to exploit and reinforce them. There is no logic that can be superimposed on the city; people make it, and it is to them, not buildings, that we must fit our plans."
— Jane Jacobs, The Death and Life of Great American Cities, 1961

i

Preface

The contemporary view of the Smart City is very much static and infrastructure-centric, focusing on installation and subsequent management of Edge devices and analytics of data provided by these devices. While this still allows a more efficient management of the city's infrastructure, optimizations and savings in different domains, the existing architectures are currently designed as single-purpose, vertically siloed solutions. This hinders active involvement of a variety of stakeholders (e.g., citizens and businesses) who naturally form part of the city's ecosystem and have an inherent interest in jointly coordinating and influencing city-level activities.

The book presents a coherent, novel vision of Smart Cities, built around a value-driven architecture. It describes the limitations of the contemporary notion of Smart City and argues that the next development step must actively include not only physical infrastructure, but ICT and human infrastructure as well. In the authors' opinion, this requires emphasizing and tightly integrating research and technological solutions from the areas of Internet of Things (IoT) and Social Computing. The book portrays the novel Smart City as an especially suitable environment for the proposed integration, and describes how the city's stakeholders would benefit from it, mostly by being able to run and participate collectively in complex, coordinated activities involving the city's infrastructure and services, but especially other stakeholders and their devices. The described activities require provisioning and scaling of the ICT infrastructure, collective communication and coordination mechanisms, as well as direct and indirect controllability mechanisms, especially with respect to human participants.

The format and the content of the book is meant to give an overview of contemporary developments in the areas of IoT and Social Computing research, and to set a research road map for a future tighter integration of the two areas in the context of the Smart City.

Vienna,
November 2016   *The authors*

Acknowledgements

The research presented in this book was supported through the following research grants and collaborations: EU FP7 600854 "SmartSociety" and Joint Programming Initiative Urban Europe, ERA-NET 5631209 "SMART-FI".

Furthermore, the authors would like to thank the dear persons without whom writing this book would not have been possible.

Schahram: *I dedicate this book to Marjan. Without you, I would not have been able to do this. With your unconditional love you supported me through the most difficult times of my life. I am also immensely thankful to my dear children, Timna and Luis, for their constant loving support and energy.*

Stefan: *I dedicate this book to you Aleksandra, as a small reciprocation for all your love and support that cannot be put in words. I also want to thank my parents Vera and Nebojsa and my sister Aleksandra for their never-ending support and wholehearted love. Finally, I would like to thank my co-authors for all the inspiring discussions and enjoyable collaboration.*

Ognjen: *I would like to thank my parents Mladenka and Milan, my brother Igor and my fiancée Katarina for their patience and support during the writing of this book. In addition, I would like to thank my co-authors for all the fruitful and creative discussions that we have had.*

Contents

Acronyms

CAS	Collective Adaptive Systems
DSL	Domain-Specific Language
H2H	Human-to-Human
HIT	Human Intelligence Task
IDE	Integrated Development Environment
IOT	Internet of Things
M2H	Machine-to-Human
M2M	Machine-to-Machine
P2P	Peer-to-Peer
PPP	Pay per Performance
PRINGL	PRogrammable INcentive Graphical Language
SCU	Social Compute Unit
SDK	Software Development Kit
SDT	Self-Determination Theory

Part I
Present and Future of Smart Cities

Preface

In absence of a single operational definition, the concept of Smart City is for many a vague notion and often subject to personal interpretation. While all existing definitions seem to agree that the ultimate goal of a Smart City is to improve the quality and the sustainability of life of its citizens through application of innovative and inclusive technology, most approaches describe the technological solutions limited to a narrow application area, falling short of presenting a comprehensive multi-dimensional perspective. This chapter, as indeed the entire book, is an attempt to present a wider perspective. It layers the high-level goal of life quality into a multi-layered architecture of values and describes the various technology interplays expected to help accomplish them.

Chapter 1
Introduction to Smart Cities and a Vision of Cyber-Human Cities

While there is no single accepted definition, the common contemporary understanding of a Smart City [75, 20] assumes a coherent urban development strategy developed and managed by city governments seeking to plan and align in the long term the management of the various city's infrastructural assets and municipal services with the sole objective of improving the quality of life for the citizens [59, 159]. The ICT role in the current Smart City vision is passive – related to collecting and analyzing data, predicting and optimizing infrastructure utilization, as well as facilitating communication between different city services and automated management of infrastructure.

Although extremely complex, the Smart City of today can perhaps best be described as a city planning/urban development methodology heavily relying on ICT to gather necessary input and make optimal engineering and planning decisions. This means that the city's strategies are planned well in advance, with big investment budgets through big infrastructural budgets. More importantly, the citizen is also put into a passive role. While the citizens are undeniably winners in this process as the beneficiaries of a more optimized and cheaper infrastructure they are not taking an active role in the development and daily management of the city.

We denominate the current stage in Smart City development as 'representative-smart', as opposed to 'collective-smart' – one of the terms we propose for describing the future vision of cyber-human smart cities involving a rich and active interplay of different stakeholders (primarily citizens, local businesses and authorities), effectively transforming the currently passive stakeholders into active ecosystem actors. Realizing such complex interplay requires a paradigm shift in how the physical infrastructure and people will be integrated and how they will interact.

At the heart of this paradigm shift lies the merging of two technology/research domains – Cyber-physical Systems and Socio-technical Systems – into the *value-driven* context of a Smart City. The presented Smart City vision diverges from the traditional, hierarchical relationship between the society and ICT, in which the stakeholders are seen as passive users who exclusively capitalize on the technological advancements. Rather, the architecture we propose puts value generation at the top of the pyramid and relies on "city capital" to fuel the generation of novel values and

© Springer International Publishing AG 2017
S. Dustdar et al., *Smart Cities*,
DOI 10.1007/978-3-319-60030-7_1

enhancement of traditional ones. This effectively transforms the role and broadens the involvement and opportunities of citizen-stakeholders, but also promotes the ICT from passive infrastructure to an active participant shaping the ecosystem.

1.1 Architecture of Values

The fundamental idea behind a collective-smart city is the inclusion of all its stake-holders (authorities, businesses, citizens and organizations) in the active management of the city. This includes not only the management of the city's infrastructure, but additionally the management of different societal and business aspects of everyday life. The scale and complexity of managing diverging individual stakeholder interests in the past was the principal reason for adopting a centralized city management model where elected representatives manage all aspects of the city's life and development. However, we believe that recent technological advances will enable us to share the so-far centralized decision-making and planning responsibilities directly with various stakeholders, allowing faster and better-tailored responses of the city to various stakeholder needs.

The key technological enabler for this process is the active and wide-scale use and interleaving of technologies and principles from the IoT and Social Computing domains in the urban city domain. These technologies form the basis level of the proposed architecture of values (Fig. 1.1). They allow the city to interact bidirectionally with the citizens in their everyday living, working and transport environments using various IoT edge devices and sensors, but also to actively engage citizens and other stakeholders to perform concrete tasks in the physical world, express opinions and preferences, and take decisions. The "city" does not need to be an active party in this interaction. It can serve as a trustworthy mediator providing the physical and digital infrastructure and accepted coordination mechanisms facilitating self-organization of citizens into transient, ad hoc teams with common goals. This synergy in turn enables creation of novel societal and business values.

Infrastructural values – This category includes and extends the benefits conventionally associated with the existing notion of Smart City – those related to the optimized management of shared (city-wide) infrastructure and resources. Traditionally, the management of such resources (e.g., transportation network and signalization, internet infrastructure, electricity grid) has been static and highly centralized. The new vision of a Smart City relies on the interplay of humans and the IoT-enabled infrastructure, enabling additional, dynamic, locally scoped infrastructural optimizations and interventions, e.g., optimization of physical and IT/digital infrastructure in domains such as computational resources, traffic or building management. Apart from existing static/planned optimizations (e.g., static synchronization of traffic lights), the dynamic optimizations of the infrastructure might include temporary traffic light regime changes when a car accident is detected.

Societal values – This novel value category arises through the direct inclusion and empowerment of citizens as key stakeholders of the city. The fact that through the use

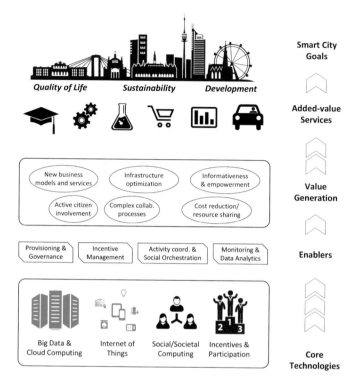

Fig. 1.1: Smart City 2.0 Architecture of values

of technology the citizens can be informed, educated, consulted and ultimately incentivized/paid to perform specific tasks in both the digital and physical environments is a powerful concept bringing along a plethora of socially significant changes.

For example, while most cities function as representative democracies, significant local changes are often decided upon through direct democracy (referendums, initiatives). While undeniably fair in principle, one of the biggest obstacles to a more frequent use of direct democracy is the underinformedness of voters [105]. It has been shown [133, 73] that informing the citizens enables them to make more judicial and responsible decisions. The pervasiveness of IoT devices enables interaction with citizens directly and opens up the possibility of informing the citizens better, or even simulating in practice the outcomes of different election choices.

Take for example the 2014 Viennese referendum where two city districts were asked to decide whether to turn one of the most frequently used shopping streets of the city into a pedestrian zone[1]. The referendum has caused much controversy, as people were skeptical that closing an important traffic artery would not cause major traffic jams. In order to give the citizens a preview of how things would work

---

[1] http://kurier.at/chronik/wien/mariahilfer-strasse-ja-fuer-umstrittene-fussgaengerzone/54.913.277 (in German)

after the transformation, the city invested in a temporary physical closing of the street and traffic re-organization the year before, as well as informational material. The total costs of the street transformation amounted to 25 million euros, out of which at least[2] 2 millions needed to be spent just to reach a common decision. In such cases, the citizens of the new Smart City can be included in the evaluation of the proposal and the decision process directly. The city can incentivize citizens to get informed about the pro et contra before making a decision; simple games and tests can raise awareness of specific problems. Interested parties can locate and engage same-minded neighbors and set up citizen collectives standing for their views. Finally, citizens can sign up to participate in cyber-physical simulations of the effects of different outcomes. For example: For turning a traffic street into a pedestrian zone, the IoT-enabled cars can be prevented from entering the street; For raising awareness of global warming, the citizens can be incentivized to have their apartments warmer/colder by a couple of degrees; To help people realize the low share of green energy, the citizens can be incentivized to use for a couple of days only the "green" percentage of the electricity they normally use. While simple, these simulations affect the citizens in their private environment through everyday (IoT) objects they interact with, and thus represent a strong motivational factor raising interest and informedness of an issue.

Business values – Apart from citizen empowerment and better inclusion in political processes, the existing research on decision making [110, 90], social orchestration and negotiation [147], and incentivization [158] provide a number of solutions for facilitating formation of collectives (groups, teams, task forces) of citizens, provisioning of necessary software support tools and digital infrastructure, algorithms for reaching agreement and compiling execution plans for different classes of tasks, as well as incentive models for both monetary and non-monetary compensation.

Combined together in the context of a Smart City, this allows the establishment of novel labor models where humans can engage in one-off or repeated activities within stationary or ad hoc created collectives, motivated by a personal interest or the offered compensation. These collaborative activities can range from the simplest on-demand crowdsourcing tasks such as deciding the color of the new subway line[3] to complex activities involving experts, such as IT incident management[163] or use of humans as sensing agents for predictive maintenance of non-IoT infrastructure, allowing for the effective and cheap inspection of local infrastructure.

Apart from offering their physical and cognitive abilities, citizens can be actively involved in enriching the Smart City infrastructure with their smart devices. The augmented infrastructure, access to the huge amounts of data and active user involvement in its maintenance can be exploited in a variety of ways, e.g., to optimize existing business models, reduce operational costs and create novel business opportunities. To be able to fully benefit from this inclusion we need novel ways to incentivize the citizens to "open source" their infrastructure, but also enable them to reap the benefits of doing so. The solution we propose lies in a combination of novel incen-

---

[2] http://kurier.at/chronik/wien/mahue-neu-es-bleibt-bei-den-25-millionen-euro-kosten/96.901.003
[3] http://qz.com/242360/stockholm-is-crowdsourcing-the-color-of-its-new-subway-line/

tive mechanisms and micro-payment technologies, which can enable fine-grained leasing and use of equipment, services and resources, as well as novel infrastructure provisioning and governance models and frameworks, which can support city-scale infrastructure management.

1.2 Smart City Platform

Contemporary Smart City development and investment strategies focus on improving the efficiency of traditional services and utilities. The focus on the "historical verticals" [80] is limiting the innovation and business potential of the city. Opening up this siloed view of the Smart City will allow more horizontal integration and creation of added values. Figure 1.2 illustrates the high-level architecture of the future Smart City Platform. The platform is a rich ecosystem that facilitates both production and consumption of added values for all the involved participants, ranging from humans to smart devices. It enables horizontal integration across different architecture layers and among different stakeholders. The main components comprising the platform include: *i) Smart City Infrastructure, ii) Core Platform Facilities*, and *iii) Value-added services*. Below, we describe these components in more detail.

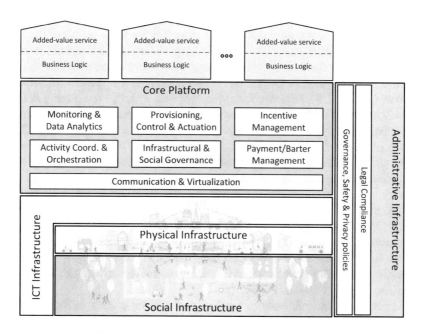

Fig. 1.2: Cyber-human Smart City platform

Starting from the *Smart City Infrastructure*, contrary to the traditionally monolithic view of a city's infrastructure, in our vision of the cyber-human Smart City, we identify different infrastructure constituents, that are inherently complementary and interdependent. *The Physical Infrastructure* consists of the union of all stakeholders' physical assets of direct interest to other stakeholders. This can include the city's transport infrastructure, electricity system, but also devices (e.g., vehicles or PV panels) owned by an individual, when they are willingly being offered to enrich the Smart City physical infrastructure and to be used by other stakeholders. *The Administrative Infrastructure* consists of the political and legal organizations governing the city's ecosystem. Collectively they act as the trusted entity determining and enforcing governance policies, guaranteeing legal and privacy protection. These organizations are not considered stakeholders of a Smart City. *The Social Infrastructure* consists of all the individual citizen and business stakeholders, i.e., of their intellectual, social and physical capabilities, as well as personal assets and resources, offered indirectly as services, individually or collectively. Examples include providing labor on a given task, or offering a ride service in a personal vehicle (as opposed to sharing the vehicle). *The ICT Infrastructure* is the cornerstone for efficient horizontal integration of different infrastructural layers and interoperability among stakeholders. It consists of all the physical and software (virtual) components for data gathering, processing, enactment of business logic, communication, and actuation of physical devices, such as sensors, IoT gateways, actuators, cloud processing and storage infrastructure and analytics software services.

Whereas the Infrastructure components resemble the vital organs of the Smart City, the *Core Platform* resembles its bloodstream, linking all the Smart City functionalities and enabling their seamless functioning. The most important functionalities of the Core Platform include: Orchestration functionalities for *Complex Coordinated Activities* (Section 1.4), *Incentive Management* (Section 1.5), *Provisioning & Governance* (Section 1.7), Monitoring & Data Analytics, as well as Control & Actuation mechanisms. Since the last two components are also present in the Smart City of today, we will not discuss them here.

The *Value-Added Services* act as the brain of the Smart City. They rely on the core platform to enable management of the Infrastructure and facilitate the value-generation process. Generally, the value-added services are largely task- and use-case-specific and we do not impose any rules or requirements on their design or functionality. They are envisioned as a playground of disruptive innovation and value generation. For example, they can be optimizations of existing business models or incubators for novel business opportunities. The value-added services are meant to follow the natural life cycle of the city's evolution and can appear and disappear in accordance with stakeholders needs.

Once in place, the value-added services can become a valuable digital asset in the ownership of the city and its citizens that can, however, extend beyond the geographical region of the city and beyond the citizens that physically reside in the city. The Republic of Estonia has recently introduced the concept of *e-residency*[4].

---

[4] https://e-estonia.com/e-residents/about/

It allows any person (non-Estonian citizen) residing outside of Estonia who fulfills specific criteria to become a legal subject (e-resident) of Estonia for a small fee. This status allows practically anyone in the world to remotely establish and manage a company based in Estonia enjoying all the benefits of modern e-Government services that the country offers, a reliable and transparent legal and e-banking system, as well as access to the entire EU market under the same conditions as any other EU company. This landmark concept, which allows Estonia to further profit from the services it offers and taxes it collects from the businesses established by the remote citizens, can equally be adopted by cities. The gain for a city from such residents is twofold – they pay fees/taxes for the use of services and at the same time they do not strain the physical infrastructure (e.g., roads, sewers). Many of the value-added services that do not rely on the physical presence of the service consumers can be offered in this fashion to external residents. The additional gains obtained in this way can help finance the transformation of a city into a Smart City and development of further services. At the same time, this concept fosters competitiveness among cities. Without the physical location/residence as the principal constraint, a citizen is free to choose any city as the provider of digital services. This means that cities will have to compete to offer better services to their citizens in an effort to keep them as residents or attract new, digital ones.

1.3 Stakeholders

We define the Smart City stakeholder as any physical or legal entity entitled by the city's authorities to use, manage and contribute to the Smart City's physical and social infrastructure. In practice this means that any citizen or business of the city, as well as city authorities or visitors who are given the right to access the Smart City platform are considered as stakeholders. Contributing to the infrastructure is equally important as using it. Stakeholders putting at the disposal their own devices, providing services and participating in collective coordinated activities (Sec. 1.4) bring in the human capital into the play, activate the infrastructure and generate novel values.

1.4 Complex Coordinated Activities

One of the principal defining characteristics of the envisioned Smart City is the existence of and support for a rich set of interactions embodied in the concept of complex, collaborative *coordinated activities*. These activities are fundamental to the generation of societal and business values described in the previous section. Whether initialized by the municipality, local businesses or the citizens themselves, a Smart City platform acts as the legal, trust and coordination enabler of such activities. On the 'physical layer', the activities comprise the following interaction types:

1. M2M – interactions between IoT devices and software services (e.g., sensing, actuation, data analytics, service compositions, micro-transactions).
2. H2H – interactions between humans/citizens (negotiation, joint planning, collaborative task execution, learning, direct democracy).
3. M2H – interactions between humans and (their) devices (notifications, personalized use, context sensing, augmented reality).

Since the machines (devices, services) are used on behalf of humans, on the more abstract level the activities represent the interactions among the various city stakeholders. In fact, the main objective of such coordinated activities is to actively facilitate the various stakeholders to (self-)organize and reach a common goal, both on a personal (micro), as well as on a city (macro) scale. The facilitation is performed through various coordination and communication mechanisms delivered by the Smart City platform. These mechanisms serve both as direct and indirect controllability methods – either enforcing specific constraints and policies (e.g., negotiation protocols, SLAs), or indirectly influencing behavioral responses of humans through incentives and peer influence. Examples of complex coordinated activities can range from collectively organized transportation [3], private infrastructure sharing[5], collective learning [60] and game-based learning[6] to gainful activities, such as collaborative software development [116]. While the size of the heterogeneous collectives participating in these activities need not be large, the potential and reach are global, allowing most citizens to participate, thus actively shaping the society, city and business environment they share.

1.5 Incentives as a Soft Controllability Principle

Managing humans in various socio-technical systems has often been criticized as neglectful of true human nature [81]. Humans are often used as role enactors in human workflows [15, 112, 6] or executors of instructions [57]. While such approaches allow the difficulties related to human-understandable context interpretation to be overcome, the human intelligence is harnessed in a passive way, since the execution is machine-driven and deterministic. This means that the collaborative and social capital of humans is not fully exploited, despite the prospective of delivering a profound positive impact on the society we live in [176]. Crowdsourcing [44] and various other platforms for collaborative consumption have partially tapped into this potential, allowing for human-driven, albeit tightly structured, collaborations.

A distinguishing characteristic of human participation in socio-technical processes is the need for motivation. Unlike software services or devices whose usage can be requested for compensation and whose outputs are deterministic, human participation is driven by personal motives, which vary individually in time and also depend on

---

[5] https://switcher.ie/broadband/news/upc-ireland-rolls-out-horizon-wi-free-service/
[6] http://www.nobelprize.org/educational/medicine/ecg/

the (social) environment. Furthermore, diverging individual motives and interests make team assembly and coordination of collective activities inherently complex.

Incentives are a means for inducing motivation and aligning disjoint individual interests in a group [155]. They include not only monetary/material rewards, but more often rely on intrinsic motivational factors, such as altruism, curiosity, competitiveness, social status. Compared to the listed role-based workflow systems where humans are issued concrete actions to perform, incentives serve as a powerful mechanism for "soft controllability" inducing wanted behavioral responses, setting psychological engagement constraints but leaving the liberty of action to the humans.

A cyber-human Smart City wishing to engage citizens into collaborative actions should offer incentive management services, such as [158], through its platform (Fig. 1.2), giving to different stakeholders the tools to motivate and engage other stakeholders into collaborative activities. The incentive management service allows the provider of the incentives to compose and tweak incentive schemes optimally for a particular purpose and a given target population. It also allows the monitoring of the incentive application and effectiveness, and subsequent adaptations. The city can incentivize the citizens to engage in decision making or to get better informed, or to change their habits (share infrastructure, promote a healthier lifestyle). Businesses aligned with such goals can provide for the costs of incentivization. Finally, where mutual resources and devices can be shared, individual citizens can set up incentive schemes to encourage bartering and partially substitute the use of money with alternative/local currencies (see Ithaca HOURS[7]) in micro-transactions having positive effects on local businesses [168] (Sec. 1.7). The incentives can be delivered through different channels, using personalized messages, to different hand-held or IoT devices. Serious games are also an attractive environment for engaging people and delivering incentives, especially for learning purposes [99]. Their captivating power is best evidenced by the recent global success of the augmented-reality game Pokemon Go[8]. As the timing and the perception of the incentive, as well as trust in the incentive provider, are the key factors of its effectiveness, we argue that the described Smart City context is a well-suited environment for the implementation of such incentive management systems. The Smart City platform provides the trusted third party technically managing the application of the incentives, while not taking an active provider role. Thanks to their pervasive distribution IoT/Edge devices are used to deliver incentive messages and provide raw data for automated monitoring of incentivized activities.

1.6 Citizen Informedness

When talking about societal values a community can most benefit from, the importance of informing, educating and actively including citizens in different life aspects

---

[7] http://ithacahours.com/

[8] http://www.bbc.com/news/technology-36763504

of the community cannot be overstated. Conventional education teaches ground truths and well-established rules, and as such will remain the fundamental growth generator for any local economy. While these ground values retain their full importance in a Cyber-Human Smart City, the new environment, being much more dynamic both in time and space, requires additional continuous learning and adaptation by the citizens living within it. Often the new knowledge is short lived and useful for a limited period or geographical space. This can prove to be especially challenging to vulnerable social groups (seniors, low-income families), which are disadvantaged when it comes to learning about the new technologies and adopting them. While this problem is already present today it is to be expected that it further escalates in a Smart City environment, where people are de-facto forced to interact increasingly with different and unknown devices built into surrounding everyday objects and varying in user interfaces and functionalities. In such cases it is important for the Smart City platform and the particular devices to offer technological support for running test trials and practical demonstrations whenever possible. Novel approaches to tackling this problem in a similar way have already started to appear (e.g., Living Labs [159]) but the research in this area has yet to get into full swing.

Apart from the need to constantly learn to adapt to an ever-changing IoT environment, citizens of our Smart City are also confronted with a sea of information which needs to be filtered to each citizen's needs and visualized in a simple format at a proper time. Some of the current technologies such as Google Now already try to implement the basic principles, by delivering highly personalized information via smart phones. In the future Cyber-Human Smart City, the spectrum of processed information will need to drastically expand to cover the usage of different IoT devices as well as past and potential interactions with other citizens and/or their devices. This also means that the information services will have to move from the purely passive aggregation and filtering functionality to interactive services which are able to dynamically gather additional information (e.g., from different users) in order to deliver only meaningful and useful information. Let us consider for example a ridesharing scenario (cf. Chapter 7). The scenario starts with a number of users submitting ride requests with offered/wanted origin-destination pairs that are then matched by the ride-sharing platform. Unless a large number of ride requests are constantly being submitted, the chances of producing enough matches are slim. This means that the matching will work satisfactorily most likely only in large and densely populated areas, and only if the ride-sharing platform is a popular application used by large numbers of users. On the other hand, if the software takes over the responsibility of monitoring various ride-sharing platforms for suitable matches and informs the user on time, the chances of producing a match rise, even outside large cities. If the initial step of submitting the ride request is instead performed autonomously by a software assistant on a user's behalf based on learned travel patterns or booking confirmations, the chances of producing a match rise further. The user, naturally, ultimately decides whether the match is useful to him, and whether to proceed with negotiations and ultimately accept the ride. In order to do this, the information service needs a runtime feedback response from the user. In the described case, the autonomous agent gathers and processes information relevant to the user at a speed that would be impossible for a human. Therefore, the informedness of the user in this case is indirect, but the actual empowerment obtained by transmitting only the relevant information is practical and concrete.

1.7 Provisioning and Governing Infrastructure as a Utility

At its core the Smart City assumes an interplay between cities and technology. At the moment this relationship is most obvious at the infrastructure level. In this regard, we mainly focus on ICT infrastructure, but due to its nature, realizing Smart City infrastructure requires a multidisciplinary effort, ranging from electrical and civil engineering and urban planing to ICT. From the ICT point of view, Smart Cities are ever stronger developing and evolving Cyber-Physical/IoT Cloud Systems that blend in Internet of Things (IoT), network elements, Cloud services and humans. This results in complex IoT Cloud infrastructures that need to be provisioned dynamically on demand and governed throughout their entire lifecycle.

The majority of traditional city infrastructure resources such as electricity or water are delivered and consumed as a public utility. Such utilities are traditionally subject to forms of public control and regulation ranging from local community-based groups to statewide government monopolies. Moreover, Smart City stakeholders engage in utility generation and consumption, as well as its distribution (e.g., sale), generally in a regulated market.

However, to date Smart City ICT infrastructure is hardly delivered and consumed as a utility. To enable this paradigm in the Smart City of the future, we identify a set of design principles that serve as a road map towards realizing the utility-based delivery and consumption of Smart City ICT infrastructure. These include: *Everything as code* – all the concerns, i.e., application business logic, but also Smart City infrastructure resources provisioning and runtime governance, should be expressed programmatically in a unified manner, as a part of the application's logic (code). *API Encapsulation* – Smart City infrastructure resources and capabilities are encapsulated in well-defined APIs, to provide a uniform view on accessing functionality and configurations of IoT cloud infrastructure. *Central point of operation* – conceptually centralized (API) interaction with Edge devices allows for a unified view of the infrastructure's provisioning and governance capabilities, without worrying about low-level infrastructure details. *Automation* – main provisioning and governance processes need to be automated in order to enable dynamic, on demand configuration and operation of the Smart City infrastructure without manual interaction with Edge devices.

Realizing the utility-based consumption of Smart City ICT infrastructure, among other things, requires rethinking traditional approaches to provisioning and governing both applications and the infrastructure. In our previous work, we have addressed some of the aforementioned challenges by introducing models and frameworks that implement and enforce some of these principles in order to facilitate utility-based provisioning and cfity-scale governance. In [120, 122], we have introduced a unified provisioning model and a framework for logically centralized provisioning large-scale, geo-distributed Smart City ICT infrastructure. This work was mainly intended to address a stringent need: To enable the Smart City ICT infrastructure to be refactored into finer-grained resource components whose behavior can be defined in software; To provide conceptually unified representation of both Edge and Cloud resources; As well as to enable automated and scalable management of IoT Cloud resources, application components and their configuration models in a logically centralized fashion. Furthermore, in [119, 125] we introduced a novel governance methodology and runtime framework for governing the Smart City infrastructure and services. The main aims here were: To bridge the current wide gap between stakeholders involved in governing Smart City systems; To enable governance strategies to be enforced in a large-scale, geographically distributed setup; and to enable dynamic, on-demand deployment and invocation of governance capabilities via cloud-based APIs.

However, although this work lays a cornerstone for realizing our vision of the Smart City, additional work needs to be done in order develop a full-fledged tool suite that is capable of facilitating the value generation chain (cf. Section 1.1). One of the key enablers is to provide novel support for realizing the delivery-consumption-compensation model for the previously introduced Smart City capital. Traditional public utilities exclusively rely on existing markets, business models and monetary institutions to realize this model. However, to realize broader participation in the previously presented architecture of values Smart Cities largely lack suitable business models for exchanging resources and services among stakeholders. Moreover, infrastructure owners and infrastructure brokers require an ecosystem to support trading Smart City services and assets.

1.8 Summary & Organization of the Book

In this chapter we have introduced a novel vision of the Cyber-Human Smart City that is based on the architecture of values. This value-driven architecture is characterized by complex coordinated activities involving the City's services, stakeholders and their smart devices. It puts the citizens in first place and promotes them to active stakeholders as opposed to passive users. We presented a set of key enablers to realize the vision of the cyber-human Smart City, which include: i) Complex Coordinated Activities, ii) Incentives as soft controllability mechanisms, iii) Citizen Informedness, and iv) Utility-based provisioning and governance of Smart City infrastructure. Finally, we presented a concrete set of design principles and requirements that serve as a manifesto of Cyber-Human Cities of the future and lay down a road map toward realizing a comprehensive Smart City platform.

This chapter both serves as a general introduction to the book and also presents a coherent vision that ties together all the components that are required to realize our vision for Smart Cities of the future. In the remainder of the book we discuss these components in depth. Part II of the book discusses our previous work related to the provisioning and governance of Smart City systems and infrastructure. In Part III, we introduce our previous work on the core technologies and technological enablers for managing the social component of the Smart City platform. Both parts (see Chapter 2 and Chapter 6), also present the state-of-the-art research and industrial efforts in the respective fields. Finally, Part III provides a road map towards Cyber-Human Smart Cities and concludes the book. We discuss the requirements and concrete technological advancements needed to move beyond contemporary Smart Cities, towards the Smart Cities of the future.

Part II
Provisioning and Governing Smart City Systems

Preface

At its core the Smart City assumes a strong interplay between cities and technology. Over recent years, cloud computing and the Internet of Things (IoT) have been converging more strongly, sparking creation of very large-scale, geographically distributed systems. These IoT Cloud systems[9] form the foundation of the ICT infrastructure of Smart Cities. In a narrow sense, Smart Cities can be viewed as complex IoT Cloud infrastructures that need to be provisioned dynamically, on demand and governed throughout their entire lifecycle. Moreover, as discussed in Chapter 1, Smart City stakeholders ever more strongly engage in utility generation and consumption, as well as its distribution (e.g., sale), generally in a regulated market. However, to date Smart City ICT infrastructure is still siloed, thus can hardly be delivered and consumed as a utility. This requires rethinking existing support for representing infrastructure resources, managing their configuration and deployment models as well as composing low-level resource components into usable infrastructures, capable of supporting novel complex coordinated activities in the context of future Smart Cities (cf. Chapter 1). In this part of the book we present our work that is mainly driven by a stringent need: To enable refactoring of the underlying infrastructure into finer-grained resource components whose behavior can be defined in software; To provide conceptually unified representation of both IoT and Cloud resources; As well as to enable automated and scalable management of Smart City infrastructures in a logically centralized fashion. In Chapter 3 we introduce a conceptual model and lay out a road map towards utility-based provisioning of Smart City infrastructures. The main building blocks of our provisioning model are software-defined IoT units. Our model conceptualizes the software-defined IoT units and elicits their main design principles together with a road map to develop corresponding technical enablers. Chapter 4 continues the line of work towards utility-based provisioning of Smart City infrastructure, by introducing middleware that provides comprehensive support for multi-level provisioning of IoT Cloud systems.

Moreover, the wide and ever more strongly growing application area of IoT Cloud in the context of Smart Cities has led to a stronger interplay and entanglement among a variety of diverse stakeholders, with different objectives, interests and backgrounds. From an application point of view IoT Cloud systems are becoming an integral enabler in optimizing urban processes, infrastructure and facilities, such as urban transportation and energy management, in order to make the cities of the future smarter and more livable. This calls for a systematic and structured approach to IoT Cloud governance. Unfortunately, contemporary Smart City governance approaches draw a hard line between high-level governance objectives (which mainly concern city representatives and business stakeholders) and operations processes. The latter concern technical stakeholders such as operations managers who need to implement concrete operations processes, conforming to or enforcing the high-level governance objectives. Therefore, at the moment there is a wide gap between the main stakeholders involved in governing Smart City applications, increasing

---

[9] Also referred to as Cyber-Physical Systems (CPS).

the risk of lost requirements or causing over-regulated systems, potentially incurring higher operation costs or limiting innovation opportunities. To address these challenges, Chapter 5 looks at Smart City governance and introduces GovOps – a novel methodology and framework for governing IoT Cloud systems. The main incentive for introducing GovOps is to bring business stakeholders and operations managers closer together and make a step forward in bridging the gap between governance objectives (e.g., standards and regulations) and operations processes. GovOps introduces a novel methodology, governance model and roles, in order to enable seamless integration and alignment of high-level governance objectives and strategies with executable operations processes from early design stages. Chapter 5 also introduces a runtime framework, which is a reference GovOps implementation, and its main purpose is to support operations managers in implementing and executing GovOps processes in large-scale Smart City systems, without worrying about scale, geographical distribution and dynamicity of such systems.

Chapter 2
State Of The Art & Related Work

2.1 Overview of Development Support for IoT Cloud Applications

Developing and managing IoT Cloud systems and applications has been receiving a lot of attention lately. In [189, 42, 69] the authors mostly deal with device virtualization and its management on cloud platforms. A number of different approaches (e.g., [167, 7]) employ semantics aspects to enable discovering, linking and orchestrating heterogeneous IoT devices. In [30, 94] the authors propose utilizing the cloud for additional computation resources – and approaches presented in [169, 190] focus on utilizing the cloud's storage resources for sensory data. Approaches presented in [38, 92] deal with integrating IoT devices and services with enterprise applications based on the SOA paradigm. These approaches mostly adopt a cloud-centric view of IoT Cloud applications development. For example, in [42] the authors focus on developing a virtualized infrastructure to enable sensing and actuating as a service in the cloud. They propose a software stack that includes support for management of device identification and device services aggregation. In [189] the authors introduce sensor-cloud infrastructure that virtualizes physical sensors on the cloud and provides management and monitoring mechanisms for the virtual sensors. Although such approaches facilitate development of IoT Cloud applications to a certain extent, they usually do not define a structured development model for such applications. This leaves many of the challenges to be resolved ad hoc when developing Smart City Applications. Another example of the cloud-centric approach is SenaaS [7]. SenaaS mostly focuses on providing a cloud semantic overlay atop physical infrastructure. It defines an IoT ontology to mediate interaction with heterogeneous devices and data formats, exposing them as event streams to the upper layer cloud services. Similarly, the OpenIoT framework [167] focuses on supporting IoT service composition by following the cloud/utility based paradigm. It mainly relies on semantic web technologies and CoAP to enable web of things and linked sensory data. To realize our vision of the Cyber-Human City (cf. Chapter 1, contemporary support for application development needs to be extended with novel programming abstractions that enable

© Springer International Publishing AG 2017
S. Dustdar et al., *Smart Cities*,
DOI 10.1007/978-3-319-60030-7_2

the everything-as-code paradigm, facilitating development of IoT Cloud applications and making the entire development process traceable and auditable (e.g., with source control systems), in order to improve maintainability and reduce development costs of Smart City applications.

Putting more focus on the edge devices, i.e., IoT gateways, network devices, cloudlets and small clouds, different approaches have emerged recently. For example, in [17] the authors present a concept of fog computing and define its main characteristics, such as location awareness, reduced latency and general QoS improvements. They focus on defining a virtualized platform that includes the edge devices and enables custom application logic to be run atop different resources throughout the network. Further, in [65] the authors focus on abstracting devices as services and enabling two-way communication between enterprise applications and devices via Web Services (WS) and provide mechanisms for service discovery and provisioning. A similar approach is DPWS [127], i.e., SOA4D or WS4D. Also, approaches utilizing RESTful protocols, CoAP [56] and sMAP [37] exist. For example, [92] focuses on defining a CoAP-based runtime to enable composition of IoT services. Most of these approaches focus on abstracting the underlying hardware and providing service-based access to a device. Although they provide some key elements, e.g., service discovery and resource management, they implicitly assume developers have a good understanding of the underlying domain, as raw sensory data streams and low-level device services are directly exposed to them and application development is envisioned by composing the atomic services into admissible control sequences or processing schemes. To enable scalable development and provisioning of Smart City applications high-level abstractions and models need to be developed in order to facilitate development of cloud-scale IoT Cloud applications for Smart Cities of the future.

Another edge-centric approach is usage of component-based frameworks [87, 13], to abstract devices or more precisely to create proxies, which are represented as components and enable remote communication with the devices. These frameworks use OSGi for component management and execution environment. However, they abstract devices as components and define a local component model and their applications operate on a residential gateway scale. The main limitation of such approaches in the context of Smart Cities is that they only provide rudimentary support for development of Smart City applications and services, which are able to seamlessly utilize both the Edge and the Cloud. In general, compared to the aforementioned edge-centric approaches our approaches presented in this part of the book also aim at better utilization of the edge infrastructure. Additionally, we also focus on providing a systematic approach, supporting application developers in addressing most of the application/infrastructure provisioning and governance issues programmatically, in a logically centralized fashion.

Another related field is macroprogramming of sensor networks [104, 31, 114, 27]. For example, in [104] the authors provide an SQL-like interface where the entire network is abstracted as a relational database (table). Similarly, in [114], the authors deal with enabling dynamic scopes in WSN, mainly addressing the important issues of task placement and data exchange (among the WSN nodes), in order to account for the heterogeneity of the nodes and enable logically localized interactions. In [31], the authors propose the notion of logical neighborhood. Their approach is based on logical nodes (templates), which enable the nodes to be instantiated and grouped, based on their exported attributes. To facilitate communication within the neighborhoods, which is of great importance in WSN, they also provide an efficient routing mechanism. In [27] the authors introduce an extensible programming framework that unifies the WSN programming abstractions in order to facilitate business process orchestration with WSN. Despite the relevant efforts to integrate provisioning and business logic (e.g., template-based customizations [31]), the main focus of the aforementioned approaches is application business logic. Compared to these approaches, in Part II of this book we address the more general problem of enabling the everything-as-code paradigm, in order to also allow for capturing provisioning and governance logic for Smart City resources and services programmatically, in a structured manner.

2.2 Provisioning Approaches in the IoT Cloud

In recent years, advancing the convergence of Edge (IoT) and Cloud computing has been receiving a lot of attention. This has resulted in a number of approaches which lay a cornerstone for realizing utility-based provisioning in the IoT Cloud. For example, different approaches deal with leveraging more powerful resources such as remote, fully fledged Clouds or smaller Cloudlets and micro data centers, which are located in the proximity (single hop away) of the Edge, to enhance resource-constrained (mobile) devices. Such approaches, also referred to as cyber-foraging systems [100], mainly focus on specific tasks such as computation offloading [35, 30, 94] or data offloading (data staging) [10, 53, 190, 169]. Although they offer valuable insights about moving cloud computing closer to the Edge, as well as about smart resource utilization, management and allocation, they mainly emphasize algorithms (e.g., solvers), energy efficiency, performance (e.g., of processing or networking) and supporting architectures for the aforementioned tasks.

Other approaches which mainly adopt a cloud-centric view mostly aim at virtualizing Edge devices, predominantly sensors and actuators, on cloud platforms. In [42] the authors focus on developing a virtualized infrastructure to enable sensing and actuating as a service in the cloud. They propose a software stack that includes support for management of device identification and device services aggregation. In [43], the same authors discuss a utility-oriented paradigm for IoT, explicitly claiming resource virtualization and abstraction as their main goal. In [189] the authors introduce sensor-cloud infrastructure that virtualizes physical sensors on the cloud and provides management and monitoring mechanisms for the virtual sensors. In [69] the authors develop an infrastructure virtualization framework for wireless sensor networks. It is based on a content-based pub/sub model for asynchronous event exchange and utilizes a custom event-matching algorithm to enable delivery of sensory events to subscribed cloud users. Also the previously described approaches SenaaS [7] and the

OpenIoT framework [167] provide some support regarding IoT Cloud provisioning. However, their support is mainly focused on high-level application-provisioning aspects such as discovering, linking and orchestrating internet-connected objects and IoT services. Finally, there are various commercial solutions such as Xively [186], Carriots [26] and ThingWorx [171], which allow users to connect their sensors to the Cloud and enable remote access to and management of such sensors. The aforementioned approaches mainly focus on providing different virtualization, device interoperability and semantic-based data integration techniques for the IoT Cloud. Therefore, such approaches conceptually underpin our provisioning middleware (discussed in Chapter 4), since virtualizing Edge devices is a main precondition towards realizing the utility-based provisioning paradigm in future Smart Cities. Although some of the above-described solutions (e.g., [189, 167, 42]) provide support for provisioning and management of virtual sensors and actuators, their support is often based on tightly coupled provisioning models, e.g., static templates. Moreover, such approaches are usually meant to support specific data-centric tasks, mostly focusing on integrating various data formats, providing data-linking solutions and supporting communication protocols. Unfortunately, this support is not sufficient to realize utility-based consumption/delivery of Smart City infrastructure because it does not address some of the crucial challenges, such as providing support for multi-level provisioning and consuming both IoT and Cloud resources as general-purpose utilities.

Putting more focus on the network virtualization, programming and management, two prominent approaches have recently appeared, namely software-defined and fog computing. Different approaches have exploited and extended software-defined concepts to facilitate utilization and management of pooled sets of shared IoT Cloud resources, e.g., software-defined storage [170] and software-defined data centers [36]. Advances in more traditional software-defined networking (SDN) [91, 88, 86] have enabled easier management and programming of intermediate network resources, e.g., routers, mostly focusing on defining the networking logic, e.g., injecting routing rules into network elements. In [17] the authors present a concept of fog computing and define its main characteristics. Although the general idea of fog computing shares similarities with our approach, there is still a number of challenges to realize its full vision [187]. Further, current advances in fog computing mainly revolve around virtualization, management and programmatic control of network elements. Network resources are an integral part of IoT Cloud infrastructures and enabling their management is of vital importance for Cyber-Human Cities. However, this is out of the scope of this book and these approaches can be seen as complementary to our own approaches presented in this part of the book.

Finally, since the utility-based provisioning paradigm originated from cloud computing, it is natural that cloud computing has provided numerous tools and frameworks to support utility-based provisioning. The relevant approaches are centered around infrastructure automation and configuration management solutions such as OpsCode Chef [132], BOSH [19] and Puppet [139] as well as deployment topology orchestration approaches such as OpenStack Heat [129], AWS CloudFormation [11] and OpenTOSCA [131]. The main reasons why these solutions cannot simply be reused in the context of IoT Cloud systems are that they mostly assume an unlimited amount of available resources; they do not account for intrinsic dependence of application business logic on underlying devices; they are usually not suited for constrained environments and they often rely on features provided only by fully fledged OSs, e.g., configuration management approaches often hand off dependency resolution to OS package managers.

2.3 IoT Cloud Governance Approaches

Recently, IoT governance has been significantly gaining in importance in the context of Smart Cities. For example, in [183] the author evaluates various aspects of IoT governance, such as privacy, security and safety, ethics etc., and defines main principles of IoT governance, e.g., legitimacy and representation, transparency and openness, and accountability. In [182], the authors deal with issues of data quality management and governance. They define a responsibility assignment matrix that comprises roles, decision areas and responsibilities and can be used to define custom governance models and strategies. Traditional IT governance approaches, such as SOA governance [12, 28, 126] and governance frameworks such as CMMI [5], the 3P model [151], and COBIT [67] provide valuable insights and models which can be applied in Smart City governance processes and are crucial to realize our vision of Cyber-Human Cities. It is important to mention that our governance solutions presented in this part of the book do not attempt to define a holistic governance approach for Smart Cities, but they lay down a necessary foundation to realizing Smart City governance. Therefore, the aforementioned approaches do not conflict conceptually with our approach and they can rather be seen as methodologies and techniques complementing our own. In Chapter 9, we discuss the relationship of our governance methodology and these approaches in more detail.

Furthermore, numerous government organizations and standardization bodies deal with IoT Cloud governance. The governance concepts have been already applied to different aspects of the Internet and there is a range of organizations such as IETF, ICANN, RIRs, ISOC, IEEE, IGF, W3C that deal with specific areas of Internet governance. The EU Commission has also created task forces, research clusters and reports that deal with governance issues in IoT [48, 49, 50]. They have identified several challenges in contemporary IoT Cloud governance, for example, the difficulty of finding a common definition of IoT governance together with the different positions of many stakeholders. Also, due to the high number and heterogeneity of technologies and devices in IoT systems, IoT governance requires even more specific solutions compared to the traditional governance solutions. Moreover, current approaches in IoT governance usually address the Internet part of the IoT, e.g., in the context of Future Internet services, while operations processes mostly deal with Things as additional resources that need to be operated. Although there are approaches that facilitate operating Edge devices (e.g., [189, 43] as we discussed in the previous section), mapping governance objectives (law, compliance, etc.) to operations processes largely remains elusive to the contemporary governance approaches. Our GovOps model (presented in Chapter 5) builds on these approaches and addresses the issue of bridging the gap between governance objectives and operations processes, by introducing the GovOps manager as a dedicated stakeholder, as well as defining the suitable GovOps reference model to support early integration of governance objectives and operations processes. For high-level business stakeholders, GovOps enables continuous analysis, verification, and improvement of governance objectives and implemented strategies using a systematic approach. Furthermore, implementing the GovOps approach enables technological advantages such as greater flexibility, reduction of time-to-delivery, improved ease of operation, and shielding operations from regulatory issues.

Chapter 3
Provisioning Smart City Infrastructure

In Chapter 1, we have introduced a novel *Architecture of Values* that serves as a conceptual framework for value generation in Smart Cities. The proposed architecture puts value generation at the top of the pyramid and relies on "city capital" to fuel the generation of novel values and enhancement of traditional ones. This effectively transforms the role and broadens the involvement and opportunities of citizen-stakeholders, but also promotes ICT from passive infrastructure to an active participant shaping the Smart City ecosystem. Contemporary Smart City development and investment strategies utilize this infrastructure in order to improve the efficiency of traditional services and utilities. However, the current focus on the "historical verticals" [80] is hindering a widespread usage of the city capital, thus limiting the innovation and business potential of the city. Opening up this siloed view of the Smart City allows for more horizontal integration and creation of added values.

At its core the Smart City represents a strong entanglement and interplay between cities and technology. The ICT Infrastructure is the cornerstone for efficient horizontal integration of different Smart City infrastructural layers and interoperability among stakeholders. It consists of all the physical and software (virtual) components for data gathering, processing, enactment of business logic, communication and actuation of physical devices, such as sensors, IoT gateways, actuators, cloud processing and storage infrastructure and analytics software services. Recently, cloud computing and the Internet of Things (IoT) have been converging ever more strongly, sparking creation of very large-scale, geographically distributed systems. We refer to these novel systems as *IoT Cloud systems*. Such systems form the foundation of ICT infrastructure of Smart Cities.

Cloud computing concepts and technologies have been intensively exploited in development and management of large-scale IoT systems, e.g., in [167, 69, 189], because theoretically, the cloud offers unlimited storage, compute and network capabilities to integrate diverse types of IoT devices and provide an elastic runtime infrastructure for IoT systems. A self-service, utility-oriented model of cloud computing can potentially offer fine-grained IoT resources in a pay-as-you-go manner, reducing upfront costs and possibly creating cross-domain application opportunities and enabling new business and usage models in Smart Cities. However, most of the

© Springer International Publishing AG 2017
S. Dustdar et al., *Smart Cities*,
DOI 10.1007/978-3-319-60030-7_3

contemporary approaches dealing with IoT Cloud systems largely focus on data and device integration by utilizing cloud computing techniques to virtualize physical sensors and actuators. Although there are approaches providing support for provisioning and management of the virtual IoT infrastructure (e.g., [189, 167, 42]), the convergence of IoT and cloud computing is still at an early stage. System designers and operations managers face numerous challenges to realize large-scale IoT cloud systems in practice, mainly because these systems impose diverse requirements in terms of granularity and flexibility of IoT resource consumption, custom provisioning of IoT capabilities such as communication protocols, elasticity concerns and runtime governance. For example, modern large-scale IoT cloud systems heavily rely on the cloud and virtualized IoT resources and capabilities (e.g., to support complex, computationally expensive analytics), thus these resources need to be accessed, configured and operated in a unified manner, with a central point of management. Further, IoT systems are envisioned to run continuously, but they can be elastically scaled in/down in off-peak times, e.g., when a demand for certain data sources is reduced. Due to the multiplicity of the involved stakeholders with diverse requirements and business models, modern IoT cloud systems increasingly need to support different and customizable usage experiences. Therefore IoT cloud systems need to support virtualization of IoT resources and IoT capabilities (e.g., gateways, sensors, data streams and communication protocols), but also enable: i) their encapsulation in a well-defined API, at different levels of abstraction, ii) A central management of configuration models and their automatic propagation to the edge of the infrastructure, iii) automated provisioning of IoT resources and IoT capabilities.

In this chapter[1], we introduce the concept of *software-defined IoT units* – a novel approach to IoT Cloud computing that encapsulates fine-grained IoT resources and IoT capabilities in a well-defined API in order to provide a unified view of accessing, configuring and operating IoT cloud systems. Our software-defined IoT units are the fundamental building blocks of software-defined IoT cloud systems. They enable consumption of IoT resources at a fine granularity and allow for policy-based configuration of IoT capabilities and runtime operation of software-defined IoT cloud systems. We present a preliminary implementation of a framework for dynamic, on-demand provisioning of software-defined IoT cloud systems. By automating the main aspects of provisioning processes and supporting centrally managed configuration models, our framework simplifies provisioning of such systems and enables flexible runtime customizations.

The rest of this chapter is structured as follows: Section 3.1 presents a motivating scenario and research challenges; Section 3.2 describes the main principles and our conceptual model of software-defined IoT systems; Section 3.3 outlines the main provisioning techniques for software-defined IoT systems; Section 3.4 introduces design and implementation of our prototype, followed by its experimental evaluation; finally, Section 3.5 concludes the chapter.

---

[1] The work presented in this chapter was originally introduced by Nastic et al. in [120].

3.1 Research Context

Urban transportation and smart buildings management are two of the most important domains of Smart Cities. In this chapter, we analyze two use cases: Fleet Management System (FMS) and Building Management System (BMS), which are derived from a real-life case study, which was conducted in collaboration with our industry partners. This section approaches the FMS and BMS systems from the perspective of Smart City operations management. It illustrates tasks that need to be performed to provision such systems and derives concrete research challenges, which the operations managers currently face when provisioning city-scale, geographically distributed IoT Cloud systems.

3.1.1 Scenarios

3.1.1.1 Provisioning FMS

FMS is a real-life IoT Cloud system responsible for managing fleets of zero-emission electric vehicles deployed worldwide in different cities. For our discussion the most important functionality of the FMS is management of the electric vehicles in different environments such as university campuses, airports and golf courses. In general, the FMS supports the involved stakeholders in remotely managing the fleet vehicles dispersed among different cities in order to optimize tasks, crucial for their respective business processes.

The FMS is an IoT cloud system comprising vehicles' on-board gateways, the network and the cloud infrastructure. The main features provided by the on-board device include: a) vehicle maintenance (fault history, battery health, crash history, and engine diagnostics), b) vehicle tracking (position, driving history, and geo-fencing), c) vehicle info (charging status, odometer, serial number, and service notification), d) set up (club-specific information, maps, and fleet information). Vehicles communicate with the cloud via 3G, GPRS or a Wi-Fi network to exchange telematic and diagnostic data. On the cloud we host different FM subsystems and services to manage the data. For example: a) *Real-time vehicle status*: location, driving direction, speed, vehicle fault alarms; b) *Remote diagnostics*: equipment status, battery health and timely maintenance reminders; c) *Remote control*: overriding on-board vehicle control system in case of emergency; d) *Fleet management*: service history and fleet usage patterns. In the following we highlight some of the FMS features, that need to be considered during system provisioning:

- The FMS subsystems and services are hosted in the cloud and heavily rely on virtualized IoT resources, e.g., vehicle gateways and their capabilities. Therefore, we need to enable encapsulation and access to IoT resources and IoT capabilities via uniform APIs.

- The FMS has different requirements regarding communication protocols. The fault alarms and events need to be pushed to the services (e.g., via MQ Telemetry Transport (MQTT) [128]), when needed a vehicle's diagnostics should be synchronously accessed via RESTfull protocols such as CoAP [56] or sMAP [37]. The remote control system requires a dedicated, secure point-to-point connection. Configuring these capabilities should be decoupled from the underlying physical infrastructure, in order to allow dynamic, fine-grained customization.
- The FMS spans multiple, geographically distributed cloud instances and IoT devices that comprise FM's virtual runtime topologies. These topologies abstract a portion of the IoT cloud infrastructure, e.g., needed by a specific subsystem, thus they should support flexible configuring to allow for on-demand provisioning.
- The FMS involves a growing number of stakeholders. Therefore, we need to accommodate the scale and geographical distribution of the current FMS offering as well as support projected growth and future customization requirements.

3.1.2 Provisioning BMS

Building Management System (BMS) is an IoT Cloud control system for buildings that enables remote monitoring and control of buildings' mechanical and electrical assets and equipment such as HVAC, lighting, elevators, plumbing and fire alarm systems. In general, it connects the Smart City buildings' assets to a cloud-based platform, which provides applications for centralized management of such assets. Some of the core features of the BMS include managing the environment temperature, CO_2 emission and humidity within a building, as well as optimizing the building's energy consumption and handling predictive maintenance. For example, the climate control services are responsible for controlling the production of heating and cooling, managing air distribution systems throughout the building, and locally controlling the air mixture to achieve the desired environment temperature. Contrary to the FMS, the BMS is less dynamic and has a smaller degree of geographical distribution. In spite of this, it is a large-scale system that supports operating several thousands of buildings throughout a city.

In general, to provision BMS operations managers perform two distinct tasks: the initial deployment and staging of devices on the one hand, and updates with varying frequency and priorities on the other hand. In our scenario the BMS provider is responsible for managing several hundreds of buildings with a variety of tenants. The managed buildings are equipped with a variety of Edge devices ranging from sensors to detect smoke and heat, to elevator and door controls, to complex cooling and heating systems. They rely on gateways, which provide constrained execution environments with limited processing, storage and memory resources to execute the device firmware and simple routines. Gateways enable the basic bundling and management of a wide variety of connected entities. Due to the current market situation and the existing lack of standards in this novel field, there exists a huge heterogeneity in terms of software environments when it comes to these gateways.

Initially all these devices need to be equiped with the necessary capabilities to enable their basic functionality. The connected sensors need to be supported, the latest firmware needs to be installed and they need to be integrated into a specific deployment structure. This is followed by long-term evolution in terms of general maintenance, changing deployments, shifting capabilities as well as updating the software environment or firmware. The second kind of updates revolve around security patches and hot fixes that need to be deployed very fast in order to ensure that the whole infrastructure stays operational. These updates are time critical since delays can cause severe security problems in the whole infrastructure. Similarly to FMS we outline the following distinct requirements in the context of BMS:

- Gateways participating in an IoT infrastructure are resource constrained in terms of their processing, memory and storage capabilities.
- Our scenario deals with large-scale deployments comprising thousands of gateways with a wide variety of different supported execution environments.
- Requirements of these gateways change over time, which makes updates necessary. These updates can either be non-time-critical or time-critical, such as security updates.
- In order to sustain operations all updates need to be efficient and fast, and, therefore, have to be performed during system runtime, without interrupting its operation, i.e., down time.

3.1.3 Research Challenges

The limited support for fine-grained provisioning at higher levels leads to tightly coupled, problem-specific IoT infrastructure components, which require difficult and tedious provisioning and configuration management tasks on multiple levels. This inherently makes provisioning and runtime operation of IoT cloud systems a complex task. Consequently, system designers and operations managers face numerous challenges to provision and operate large-scale IoT cloud systems such as the FMS or BMS.

RC1 – The IoT cloud services and subsystems provide different functionality or analytics, but they mostly rely on common physical IoT infrastructure. However, to date the IoT infrastructure resources have been mostly provided as coarse-grained, rigid packages, in the sense that the IoT systems, e.g., the infrastructure components and software libraries, are specifically tailored for the problem at hand and do not allow for flexible customization and provisioning of the individual resource components or the runtime topologies.

RC2 – *Elasticity*, although one of the fundamental traits of traditional cloud computing, has not yet received enough attention in IoT cloud systems. Elasticity is a principle of provisioning the required resources dynamically and on demand, enabling applications to respond to varying load patterns by adjusting the amount of provisioned resources to exactly match their current needs, thus minimizing resource over-provisioning and allowing for better utilization of the available resources [47].

However, IoT cloud systems are usually not tailored to incorporate elasticity aspects. For example, new types of resources, e.g., data streams, delivered by IoT infrastructure are still not provided elastically in IoT cloud systems. Opportunistic exploitation of constrained resources, inherent to many IoT cloud systems, further intensifies the need to provision the required resources on demand or as they become available. These challenges prevent current IoT systems from fully utilizing the benefits the cloud's elastic nature has to offer and call for new approaches to incorporate the elasticity capabilities in IoT cloud systems.

RC3 – Dependability is a general measure of dynamic system properties, such as availability, reliability, fault resilience and maintainability. Cloud computing supports development and operation of dependable large-scale systems atop commodity infrastructure, by offering an abundance of virtualized resources, providing replicated storage, enabling distributed computation with different availability zones and diverse, redundant network links among the system components. However, the challenges to build and *operate dependable large-scale IoT cloud systems* are significantly aggravated because in such systems the cloud, network and embedded devices converge, thus creating very large-scale hyper-distributed systems, which impose new concerns that are inherently elusive with traditional operations approaches.

RC4 – Due to dynamicity, heterogeneity, geographical distribution and the sheer scale of the IoT cloud, traditional management and provisioning approaches are hardly feasible in practice. This is mostly because they implicitly make assumptions such as physical on-site presence, manually logging into devices, understanding a device's specifics etc., which are difficult, if not impossible, to achieve in IoT cloud systems. Thus, novel techniques, to provide unified and conceptually centralized view of a system's configuration management are needed.

Therefore, we need novel models and techniques to provision and operate IoT cloud systems, at runtime. Some of the obvious requirements to make this feasible in the very large-scale, geographically distributed setup are: (i) We need tools which will automate development, provisioning and operations (DevOps) processes; (ii) Supporting mechanisms need to be late-bound and dynamically configurable, e.g., via policies; (iii) Configuration models need to be centrally managed and automatically propagated to the edge of the infrastructure; (iv) Processes such as configuration models enforcement and deployment need to be flexibly repeatable with as little effort as possible.

3.2 Main Building Blocks of Software-Defined IoT Systems

3.2.1 Design Principles of Software-Defined IoT Cloud Systems

Generally, software-defined denotes the principle of abstracting the low-level components, e.g., hardware, and enabling their provisioning and management through a well-defined API [96]. This enables refactoring the underlying infrastructure into

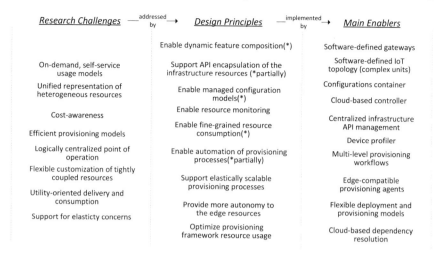

Fig. 3.1: Summary of main principles and enablers of software-defined IoT Cloud systems

finer-grained resource components whose functionality can be defined in software after they have been deployed.

Software-defined IoT Cloud systems comprise a set of resource components, hosted in IoT Cloud, which can be provisioned and controlled at runtime. The IoT resources (e.g., sensory data streams), their runtime environments (e.g., gateways) and capabilities (e.g., communication protocols, analytics and data point controllers) are described as *software-defined IoT units*. *Software-defined IoT units* are software-defined entities that are hosted in an IoT cloud platform and abstract accessing and operating underlying IoT resources and lower-level functionality. Generally, *software-defined IoT units* are used to encapsulate IoT Cloud resources and lower-level functionality and abstract their provisioning and governance, at runtime. To this end, our *software-defined IoT units* expose well-defined APIs and they can be composed at different levels, creating virtual runtime topologies on which we can deploy and execute IoT cloud systems such as our FM system. The main design principles of software-defined IoT Cloud systems that we discuss in this chapter are marked with "*" in Figure 3.1 and are described in more detail subsequently. Other design principles, shown in the same figure, are discussed later in Chapter 4.

- API Encapsulation – IoT resources and IoT capabilities are encapsulated in well-defined APIs, to provide a unified view of accessing functionality and configurations of IoT cloud systems.
- Fine-grained consumption – The IoT resources and capabilities need to be accessible at different granularity levels to support agile utilization and self-service consumption.

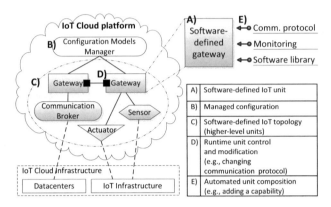

Fig. 3.2: Main enablers of software-defined IoT cloud systems

- Enable dynamic feature composition – The units are specified declaratively and their functionality is defined (composed) programmatically in software, using the well-defined API and available, familiar software libraries.
- Automated provisioning – Main provisioning processes need to be automated in order to enable dynamic, on-demand configuring and operating software-defined IoT systems, on a large scale (e.g., hundreds gateways).
- Managed configuration models – The configuration models need to be managed automatically, as well as dynamically propagated and (re)enforced in the edge resources, by a provisioning framework.

Figure 3.1 summarizes how we translate the aforementioned high-level design principles into concrete technical enablers. It serves as a general road map towards achieving our goal of enabling the utility-based provisioning paradigm in IoT Cloud systems. For example, to allow for flexible system customization, we need to enable fine-grained resource consumption and well-defined API encapsulation and provide support for policy-based specification and configuration. Among other things, these principles are enabled by our software-defined IoT units and support for centrally managed configuration models. Figure 3.2 gives a high-level graphical overview of the main building blocks and enabling techniques, which are the prime focus of this chapter. Subsequently, we describe them in more detail. In Chapter 4, we will focus on enabling the remaining design principles shown in Figure 3.1.

3.2.2 Conceptual Model of Software-Defined IoT Units

Figure 3.3 illustrates the conceptual model of our software-defined IoT units. The units encapsulate functional aspects (e.g., communication capabilities or sensor poll frequencies) and non-functional aspects (e.g., quality attributes, elasticity capabilities, costs and ownership information) of the IoT resources and expose them in the IoT

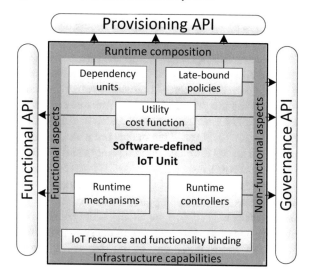

Fig. 3.3: Conceptual model of software-defined IoT units

cloud. The functional, provisioning and governance capabilities of the units are exposed via *well-defined APIs*, which enable provisioning and control of the units at runtime, e.g., start/stop. Our conceptual model also allows for composition and interconnection of software-defined IoT units, in order to dynamically deliver IoT resources and capabilities to the applications. The runtime provisioning and configuration is performed by specifying late-bound policies and configuration models. Naturally, the software-defined IoT units support mechanisms to map the virtual resources to the underlying physical infrastructure.

To technically realize our unit model we introduce a concept of *unit prototypes*. They can be seen as resource containers, which are used to bootstrap more complex, higher-level units. Generally, they are hosted in the cloud and enriched with functional, provisioning and governance capabilities, which are exposed via software-defined APIs. The unit prototypes can be based on OS-level virtualization, e.g., VMs, or finer-grained kernel-supported virtualization, e.g., Linux containers. Conceptually, virtualization choices do not impose any limitations, because by utilizing the well-defined API, our unit prototypes can be dynamically configured, provisioned, interconnected, deployed and controlled at runtime.

Given our conceptual model (Figure 3.3), by utilizing the *provisioning API*, the unit prototypes can be dynamically coupled with late-bound runtime mechanisms. These can be any software components (custom or stock), libraries or clients that can be configured and whose binding with the unit prototypes is differed to the runtime. For example, the mechanisms can be used to dynamically add communication capabilities, new functionality or storage to our software-defined IoT units. Therefore, by specifying policies that are bound later during runtime, system designers or operations managers can flexibly manage unit configurations and customize their capabilities, at *fine granularity levels*. Our conceptual model also allows for composing the software-defined IoT units at higher levels. By selecting dependency units, e.g., based on their costs, analytics or elasticity capabilities, and linking them together, we can dynamically build more complex units. This enables flexible *policy-based specification and configuration* of complex relationships between the units. Therefore, by carefully choosing the granularity of our units and providing configuration policies we can *automate the unit composition process* at different levels and in some cases completely defer it to runtime. This makes the provisioning process flexible, traceable and repeatable across different cloud instances and IoT infrastructures, thus reducing time, errors and costs.

The runtime *governance API*, exposed by the units, enables us to perform runtime control operations such as starting or stopping the unit or to change the topological structure of the dependency units, e.g., dynamically adding or removing dependencies at runtime. Therefore, one of the most important consequences of having software-defined IoT units is that the functionality of the virtual IoT infrastructure can be (re)defined and customized after it has been deployed. New features can be added to the units and the topological structure of the dependency units can be customized at runtime. This enables automation of provisioning and governance processes, e.g., by utilizing the governance API and providing monitoring at unit level, we can enable *elastic horizontal scaling* of our units. Therefore, the most important features of software-defined IoT units which enable the general principles of software-defined IoT (see Section 3.2.1) are: i) They provide software-defined API, which can be used to access, configure and control the units, in a unified manner. ii) They support fine-grained internal configurations, e.g., adding functional capabilities such as different communication protocols, at runtime. iii) They can be composed at a higher-level, via dependency units, creating virtual topologies that can be (re)configured at runtime. iv) They enable decoupled and managed configuration (via late-bound policies) to provision the units dynamically and on demand. v) They have utility cost functions that enable IoT resources to be priced as utilities.

3.2.3 Unit Classification

Depending on their purpose and capabilities, our software-defined IoT units have different granularity and internal topological structure. Therefore, conceptually we classify them into: (i) *atomic*, (ii) *composed* and (iii) *complex software-defined IoT units*. Depending on their type, the units require specific runtime mechanisms and expose specific provisioning API. Figure 3.4 depicts a simplified model of the software-defined IoT unit structure and the most important dependencies among the described unit types.

The *atomic software-defined IoT units* are the finest-grained software-defined IoT units, which are used to abstract the core capabilities of an IoT resource. They provide software-defined API and need to be packaged portably to include components and libraries, that are needed to provide desired capabilities. Figure 3.5 depicts some

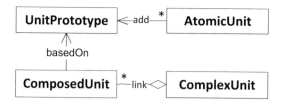

Fig. 3.4: Simplified model of software-defined IoT units structure

examples of the atomic software-defined units. We broadly classify them into functional and non-functional atomic software-defined IoT units, based on the capabilities they provide. Functional units encapsulate capabilities such as communication or IoT compute and storage. Non-functional units encapsulate configuration models and capabilities such as elasticity controllers or data-quality enforcement mechanisms. Therefore, the atomic units are used to identify fine-grained capabilities needed by an application. For example, the application might require communication to be performed via a specific transport protocol, e.g., MQTT, or it might need a specific monitoring component, e.g., Ganglia[2]. Classifications similar to the one presented in Figure 3.5 can be used to guide the atomic units selection process, in order to easily identify the exact capabilities needed by the application.

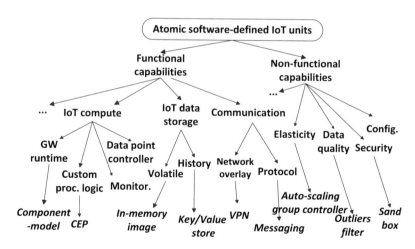

Fig. 3.5: Example classification of atomic software-defined IoT units

The *composed software-defined IoT units* have multiple functional and non-functional capabilities, i.e., they are composed of multiple atomic units. Similarly to the atomic units they provide well-defined API, but require additional functionality such as

---

[2] http://ganglia.info/

mechanisms to support declaratively composing and binding the atomic units, at runtime (Section 3.3.2). An example of a composed unit is a software-defined IoT gateway.

The *complex software-defined IoT units* enable capturing complex relationships among the finer-grained units. Internally, they are represented as a topological network, which can be configured and deployed, e.g., in the cloud. They define an API and can integrate (standalone) runtime controllers to dynamically (re)configure the internal topology, e.g., to enable elastic horizontal scaling of the units. Finally, they rely on runtime mechanisms to manage references, e.g., IP addresses and ports, among the dependency units.

We notice that the software-defined API and our units offer different advantages to the stakeholders involved in designing, provisioning and governing software-defined IoT systems. For example, IoT infrastructure providers can offer their resources at fine granularity, on demand. This enables specifying flexible pricing and cost models and allows for offering the IoT resources as elastic utilities in a pay-as-you-go manner. Because our units support *dynamic and automated composition* on multiple levels, consumers of IoT cloud resources can provision the units to exactly match their functional and non-functional requirements, while still taking advantage of the existing systems and libraries. Further, system designers and operations managers, use late-bound policies to specify and configure the unit's capabilities. Because we treat the functional and configuration units in a similar manner (see Section 3.3.2), configuration models can be stored, reused, modified at runtime and even shared among different stakeholders. This means that we can support *managed configuration models*, which can be centrally maintained via configuration management solutions for the IoT cloud, e.g., based on OpsCode Chef[3], Bosh[4] or Puppet[5].

3.3 Main Techniques for Provisioning Software-Defined IoT Cloud Systems

3.3.1 Automated Composition of Software-Defined IoT Units

Generally, building and deploying software-defined IoT cloud systems includes creating and/or selecting suitable software-defined IoT units, configuring and composing more complex units and building custom business logic components. The deployment phase includes deploying the software-defined IoT units together with their dependency units and required (possibly standalone) runtime mechanisms (e.g., a message broker). In this chapter we mostly focus on provisioning reusable stock components such as gateway runtime environments or available communication protocols.

---

[3] http://opscode.com/chef

[4] http://docs.cloudfoundry.org/bosh/

[5] http://puppet.com

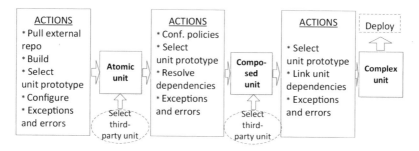

Fig. 3.6: Automated composition of software-defined IoT units

Figure 3.6 illustrates the most important steps in composing and deploying our IoT units. There are three levels of configuration that can be performed: (i) Building/selecting atomic units; (ii) Configuring composed units; (iii) Linking into complex units. Each of the phases includes selecting and provisioning suitable unit prototypes. For example, the unit prototypes can be based on different resource containers such as VMs, Linux Containers (e.g., Docker) or OSGi runtime.

The atomic units are usually provided as stock components, e.g., by a third party, possibly in a market-like fashion. Therefore, this phase usually involves selecting and configuring stock components (e.g., Sedona[6] or Niagara [AX7] execution environments). Classifications similar to the one presented in Figure 3.5 can be used to guide the atomic unit selection process. In case we want to perform custom builds of the existing libraries and frameworks, there are many established build tools that can be used, e.g., for Java-based components, Apache Ant or Maven.

On the second level, we configure the composed units, e.g., a software-defined IoT gateway. This is done by adding the atomic units (e.g., runtime mechanisms and/or software libraries) to the composed unit. For example, we might want to enable the gateway to communicate over a specific transport protocol, e.g., MQTT, and add a monitoring component to it, e.g., a Ganglia agent. To perform this composition seamlessly at runtime, additional mechanisms are required. We describe them in Section 3.3.2.

The third level includes defining the dependency references between the composed units, which "glue together" the complex units. These links specify the topological structure of the desired complex units. For example, to this end we can set up a virtual private network and provide each unit with a list of IP addresses of the dependency units. In this phase, we can use frameworks (e.g., TOSCA-based, OpenStack Heat, Amazon CloudFormation etc.) to specify the runtime topological structure of our units and utilize mechanisms (e.g., Ubuntu CloudInit[8]) to bootstrap the composition.

---

[6] http://www.sedonadev.org/

[7] http://www.niagaraax.com/

[8] http://help.ubuntu.com/community/CloudInit/

3.3.2 Centrally Managed Configuration Models

An important concept behind software-defined IoT cloud systems is that of late-bound runtime policies. Our units are configured declaratively via policies by utilizing the exposed software-defined API, without worrying about the internals of the runtime mechanisms, i.e., the atomic units. To enable seamless binding of the atomic units we provide a special unit prototype, called the *bootstrap container*. The bootstrap container provides mechanisms to define (bind) the units based on supplied configurations or to redefine them when configuration policies are changed. Therefore, the units can be simply "dropped in" and our bootstrap container (re)binds them together at runtime without rebooting system. Therefore, in order to support centrally managed configuration models and dynamic feature composition, besides managing the units our provisioning framework is responsible for maintaining application-specific configurations. Application configuration models are treated as special components of artifact packages. By decoupling the configuration models from the functional artifacts, we can treat them like any software-defined IoT unit that adheres to the general principles of software-defined IoT (Section 3.2.1). Our framework provides mechanisms to specify and propagate the configuration models to the edge of the IoT cloud infrastructure (e.g., gateways) and our bootstrap container enforces the provided directives.

To support fully fledged dynamic feature composition, the configuration container can act as a plug-in system, based on the inversion of control principles. It provides mechanisms to bind the application artifacts (e.g., atomic units) based on supplied configurations or to redefine them when configurations are changed. The container initially binds such functional artifacts based on the configuration models and continuously listens for configuration changes, applying them on the affected functional artifacts accordingly. Runtime changes are achieved by invalidating affected parts of the existing dependency tree and dynamically rebuilding them, based on the new configuration directives. This feature is especially useful for managing the communication protocols, which are provided by cloud and device connectivity components (cf. Chapter 4). However, to support dynamic feature composition, our framework requires the artifacts to be wrapped in well-defined APIs, which are known to the provisioning container. Since this imposes some limitations, this feature is optionally provided by the framework. The main advantage of this approach is that it enables updating configuration models without updating the entire artifact package, thus allowing for flexible customizations and dynamic configuration changes without runtime interrupts as well as reducing communication overhead.

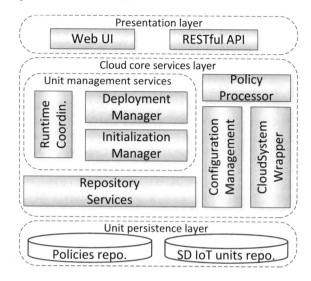

Fig. 3.7: Framework architecture overview

3.4 Prototype Implementation & Evaluation

3.4.1 Preliminary Implementation of Provisioning Controller

The main aim of our prototype is to enable developers and operations managers to dynamically, on-demand provision and deploy software-defined IoT systems. This includes providing software-defined IoT unit prototypes, enabling automated unit composition at multiple levels and supporting centralized runtime management of the configuration models.

In Section 3.2 we introduced the conceptual model of our software-defined IoT units. To technically realize our units, we utilize the concept of virtual resource containers. More precisely, we provide different *unit prototypes* that can be customized and/or modified at runtime by adding required runtime mechanisms encapsulated in our atomic units. The unit prototypes provide resources with different granularity, e.g., VM flavors, group quotas, priorities etc., and boilerplate functionality to enable automated provisioning of custom software-defined IoT units. Figure 3.7 provides a high-level overview of the framework (cloud-based provisioning controller) architecture. Our framework is completely hosted in the cloud and follows a modular design that guarantees flexible and evolvable architecture. The current prototype is implemented atop OpenStack [130], which is an open source Infrastructure-as-a-Service (IaaS) cloud computing platform. *The Presentation layer* provides user interface via a Web-based UI and RESTful API. They allow a user to specify various configuration models and policies, which are used by the framework to compose and deploy our units in the cloud. *Cloud core services layer* contains the main functionality of the framework. It includes the *PolicyProcessor* used to read the input configurations= and transform it into the internal model defined in our framework. *Units management services* utilize this model for composing and managing the units. The *InitializationManager* is responsible for configuring and composing more complex units. It translates the directives specified in configuration models into concrete initialization actions on the unit level. In our current implementation, the core of the *InitalizationManager* is an OpsCode Chef client, which is passed to the VMs during initialization via Ubuntu cloud-init. *InitalizationManager* also provides mechanisms for configuration management. The *DeploymentManager* is used to deploy the software-defined IoT units in the cloud. Our prototype relies on SALSA[9], a deployment automation framework developed in our department. It utilizes the API exposed by the *CloudSystemWrapper* to enable deployment across various cloud providers, currently implemented for OpenStack cloud. The *DeploymentManager* is responsible for managing and distributing the dependency references for the complex units (Section 3.2.3). The *Unit persistence layer* provides functionality to store and manage our software-defined units and policies.

3.4.2 Experiments

3.4.2.1 Revisiting the Motivating Scenario

We now show how our prototype is used to provision a complex software-defined IoT unit, which provides functionality for the real-life FMS location-tracking service (Section 3.1.1). The service reports vehicle location in near real-time in the cloud. To enable remote access, the monitored vehicles have an on-board device, acting as a gateway to its data and control points. To improve performance and reliability, the golf course provides on-site gateways, which communicate with the vehicles, provide additional processing and storage capabilities and feed the data into the cloud. Therefore, the physical IoT infrastructure comprises network-connected vehicles, on-board devices and local gateways.

Typically, to provision the FMS service system designers and operations manager would need to directly interact with the rigid physical IoT infrastructure. Therefore, they at least need to be aware of its topological structure and devices' capabilities. This means that the FMS service also needs to have understanding of the IoT infrastructure, instead of being able to customize the infrastructure to its needs. Due to the inherent inflexibility of IoT infrastructure, its provisioning usually involves long and tedious tasks such as manually logging into individual gateways, understanding gateway internals or even on-site presence. Therefore, provisioning even a simple FMS location-tracking service involves performing many complex tasks. Due to a large number of geographically distributed vehicles and involved stakeholders IoT infrastructure provisioning requires a substantial effort prolonging service delivery

---

[9] https://github.com/tuwiendsg/SALSA/

and increasing costs. Subsequently, we show the advantages our units (Section 3.2.2) and provisioning techniques (Section 3.3) have to offer to operations managers and application designers in terms of: a) *Simplified provisioning* to reduce time, costs and possible errors; b) *Flexibility* to customize and modify the IoT units and their runtime topologies.

To enable the FMS system we developed a number of atomic software-defined IoT units[10] such as a software-defined sensor that reports vehicle location in real-time, messaging infrastructure based on Apache ActiveMQ[11], a software-defined protocol based on MQTT and JSON, the bootstrap container based on the Spring framework[12], and corresponding configuration units. The experiments are simulated on our OpenStack (Folsom) cloud and we use Ubuntu 12.10 cloud image (Memory: 2 GB, VCPUs: 1, Storage: 20 GB). To display location changes we develop a Web application that displays changes of vehicles' location on Google Maps.

3.4.2.2 Simplified Provisioning

To demonstrate how our approach simplifies provisioning of the virtual IoT infrastructure, we show how a user composes the FMS complex software-defined IoT unit, using our framework. Figure 3.8 shows the custom deployment of the topological structure of the FMS vehicle-tracking unit, deployed in the cloud. The unit contains two gateways for the vehicles it tracks, a web server for the Web application and a message broker that connects them.

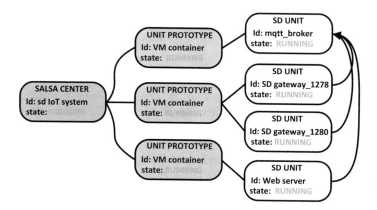

Fig. 3.8: Topological structure of FMS vehicle-tracking unit (a screen shot)

---

[10] https://github.com/tuwiendsg/SDM

[11] http://activemq.apache.org/

[12] http://projects.spring.io/spring-framework/

In order to start provisioning the complex unit, the system designer only needs to provide a policy describing the high-level resources and capabilities required by the FMS service. For example, Listing 3.1 shows a snippet from the configuration policy for the FMS location-tracking unit, which illustrates how we specify a software-defined gateway, for the on-board device.

```
1  ...
2  <tosca:NodeTemplate id="SD-Gateway"
3    name="car_1278" type="vm">
4    <tosca:Properties>
5     <MappingProperties>
6      <MappingProperty type="vm">
7       <property name="instanceType">m1.small</property>
8       <property name="provider">openstack@dsg</property>
9       <property name="baseImage">ami-00000163</property>
10      </MappingProperty>
11     </MappingProperties>
12    </tosca:Properties>
13    <tosca:Requirements>
14     <tosca:Requirement name="MQTT-broker-IP" type="String"
15     id="brokerIp_Requirement"/>
16    </tosca:Requirements>
17    <tosca:DeploymentArtifacts>
18     <tosca:DeploymentArtifact artifactType="chef"
19     artifactRef="deployClient"/>
20    </tosca:DeploymentArtifacts>
21  </tosca:NodeTemplate>
22  ...
```

Listing 3.1: Partial TOSCA-like complex unit description

The policy describes the gateway's initial configuration and the cloud instance where it should be deployed. Additionally, it defines a dependency unit, i.e., the MQTT broker, and specifies the vehicle's Id, which can be used to map it on the underlying device. Our framework takes the provided policy, spawns the required unit prototypes and provides them with references to the dependency units. At this stage the virtual infrastructure comprises solely of unit prototypes (VM-based). After performing the high-level unit composition and establishing the dependencies between the units, the user continues composing on the finer-granularity level. By applying the top-down approach we enable differing design decisions and enable early automation of known functionality, to avoid over-engineering and provisioning redundant resources.

In the next phase, the user provisions individual unit prototypes. To this end, he provides policies specifying desired finer-grained capabilities. Listing 3.2 shows example capabilities that can be added to the gateway. To enable asynchronous pushing of the location changes it should communicate over the MQTT protocol. Listing 3.3 shows part of the Chef recipe used to add the MQTT client to the gateway.

Our framework fetches the atomic units that encapsulate the required capabilities from the repository and composes them automatically, relying on the software-defined API and our bootstrap container.

```
1 {"run_list":
2   ["recipe[bootstrap_container]",
3    "recipe[mqtt-client]",
4    "recipe[protocol-config-unit]",
5    "recipe[sd-sensor]"]
6 }
```

Listing 3.2: Run list for software-defined gateway

```
1 include_recipe 'bootstrap_container::default'
2 remote_file "mqtt-client-0.0.1-SNAPSHOT.jar" do
3   source "http://128.130.172.215/salsa/upload/files/..."
4   group "root"
5   mode 00644
6   action :create_if_missing
7 end
```

Listing 3.3: Chef recipe for adding MQTT protocol

Therefore, compared to the traditional approaches, which require gateway-specific knowledge, using proprietary API, manually logging into the gateways to set data points, our *automated unit composition* (Section 3.3.1) based on declarative unit configuration policies simplifies the provisioning process and makes it traceable and repeatable. Our units can easily be shared among the stakeholders and composed to provide custom functionality. This enables system designers and operations managers to rely on existing, established systems, thus reducing provisioning time, potential errors and costs.

3.4.2.3 Flexible Customization

To exemplify the flexibility of our approach let us assume that we need to change the configuration of the FMS unit to use CoAP instead of MQTT. This can be due to requirements change (Section 3.1.1) or reduced network connectivity or simply to reuse the unit for a golf course with different networking capabilities. To customize the existing unit, an operations manager only needs to change the recipe[protocol-config-unit] unit (Listing 3.2) and provide an atomic unit for the CoAP client. This is a nice consequence of our late-bound runtime mechanisms and support for *managed configuration models* provided by our framework. We treat both functional and configuration units in the same manner and our bootstrap container manages their runtime binding (Section 3.3.2). Compared to traditional approaches that require each gateway to be addressed individually, firmware updates or even modifications on the hardware level, our framework enables flexible runtime customization of our units and supports operation managers in seamlessly enforcing the configuration baseline and its modifications on a large-scale.

3.5 Summary

In this chapter, we introduced the conceptual model of software-defined IoT units. To the best of our knowledge this is the first attempt to apply software-defined principles to IoT systems. We showed how they are used to abstract IoT resources and capabilities in the cloud, by encapsulating them in software-defined APIs. We presented automated unit composition and managed configuration, the main techniques for provisioning software-defined IoT systems. The initial results are promising in the sense that software-defined IoT systems enable sharing of the common IoT infrastructure among multiple stakeholders and offer advantages to IoT cloud system designers and operations managers in terms of simplified, on-demand provisioning and flexible customization. Therefore, we believe that software-defined IoT systems can significantly contribute to the evolution of IoT cloud systems.

Chapter 4
Middleware for Utility-based Provisioning of Smart City Infrastructure

Today, a large number of a city's facilities are organized as public utilities. Electricity, water, public transportation and various energy resources are some examples of public utilities that are delivered and managed by one party (e.g., the municipality) and that can be bought, metered and consumed by different Smart City stakeholders such as citizens. These utilities form the core infrastructure of contemporary cities and represent a significant portion of the "city capital" used to fuel a variety of city activities and processes. One of the main advantages of the utility-based consumption model is reflected in its support for self-service, on-demand resource consumption, where users can dynamically consume (allocate) the appropriate amount of infrastructure resources. For example, in case of ICT infrastructure this can be compute or storage resources required by an application or value-added service [24, 8]. As we have seen in Chapter 3, one of the main traits of future Smart Cities is that ICT infrastructure and data are ever more strongly becoming a crucial part of the city capital, enabling generation of novel values in the Smart City.

Unfortunately, realizing the utility-based provisioning paradigm of Smart City ICT infrastructures is still in its infancy, mainly because current approaches dealing with IoT Cloud[1] provisioning focus on providing virtualization solutions for the IoT devices, such as IoT gateways [43, 189, 167]. Although device virtualization is one of the preconditions for utility-based provisioning, such approaches usually focus on vertical solutions (cf. Chapter 1). They are intended to support a specific task, e.g., data integration or data linking, and largely rely on rigid provisioning models. This inherently prevents the consumption of Smart City infrastructure resources as generic utilities and requires rethinking existing support for: i) representing the Smart City infrastructure resources, ii) managing their delivery, configuration, consumption and pricing models, as well as iii) composing low-level resource components into usable infrastructures, capable to support novel Complex Coordinated Activities in the context of future Smart Cities (cf. Chapter 1).

---

[1] The IoT Cloud is the core of Smart City ICT infrastructure, as we have thoroughly discussed in Chapter 3.

© Springer International Publishing AG 2017
S. Dustdar et al., *Smart Cities*,
DOI 10.1007/978-3-319-60030-7_4

In this chapter[2], we continue our line of research towards utility-based provisioning of Smart City infrastructures and introduce a novel provisioning middleware for the IoT Cloud. Our middleware builds on the previously introduced concepts and frameworks of Chapter 3, extending them with comprehensive support for scalable multi-level provisioning of IoT Cloud systems. This is one of the crucial preconditions for realizing the utility-based provisioning paradigm in Smart Cities and IoT Cloud systems. The main features of our middleware include: i) Support for automated provisioning of infrastructure resources, application components and configuration models in a uniform, logically centralized manner through dynamically managed APIs; ii) Extensible and flexible provisioning models, which support self-service, on-demand consumption of Edge-device resources; iii) A generic, light weight resource abstraction mechanism, which allows for application-specific customizations of and virtually exclusive access to low-level devices, e.g., sensors and actuators, with well-defined APIs.

The remainder of the chapter is organized as follows: Section 4.1 presents the main research challenges and the research context; In Section 4.2 we introduce our middleware and discuss its architecture in detail; Section 4.3 outlines the major runtime mechanisms for multi-level provisioning; Section 4.4 describes experimental results and outlines the current prototype implementation; Finally, Section 4.5 concludes the chapter and gives an outlook of our future research towards realizing fully fledged utility-based provisioning of Smart City infrastructure.

4.1 Research Context

In Chapter 3, we have introduced a Smart City infrastructure-provisioning model based on software-defined IoT Cloud systems. The core concept of the provisioning model is *software-defined IoT units*. They describe IoT Cloud resources (e.g., virtual sensors), their runtime environments (e.g., gateways) and capabilities (e.g., communication protocols or data point controllers). Such units are used to encapsulate the IoT Cloud resources and abstract their provisioning in software. To this end, they expose well-defined APIs and can be composed at different levels, creating virtual runtime infrastructures for IoT Cloud applications.

The main purpose of such software-defined IoT Cloud infrastructures is to enable *utility-based provisioning of IoT Cloud resources* by providing a uniform and logically centralized view of the entire underlying resource pool, as well as by allowing IoT Cloud applications to customize and consume those resources dynamically and on demand. However, due to the dynamicity, heterogeneity, geographical distribution and sheer scale of such infrastructures, achieving these features poses a number of challenges. To better motivate our work, in the following we discuss the properties of IoT Cloud infrastructures and derive a set of key research challenges that currently prevent utility-based provisioning of IoT Cloud resources.

---

[2] The work presented in this chapter was originally introduced by Nastic et al. in [123].

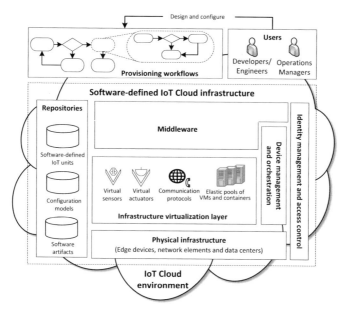

Fig. 4.1: Overview of software-defined IoT Cloud infrastructure

Figure 4.1 depicts a high-level architecture overview of the software-defined infrastructure, and shows how the main stakeholders interact with such infrastructure. The bottom layer represents the *Physical infrastructure*, which comprises a variety of geographically dispersed edge devices (e.g., sensors and gateways), network elements (routers and switches) and large data centers. In reality, the physical infrastructure is usually not flat and follows a hierarchical structure, where sensors and actuators are connected to data centers via gateways, which are intermediary nodes that mediate the communication, but also provide constrained computational and storage resources, which are currently largely underutilized. Additionally, it is common to strategically place more powerful processing nodes near the Edge (but within the hierarchy), such as Cloudlets and micro data centers. The communication between the Edge and the data centers is realized over heterogeneous networks which include wired, wireless and cellular communication channels. Moreover, IoT Cloud infrastructure is highly decentralized and distributed among multiple geographical regions and organizations.

A distinguishing feature of the software-defined IoT Cloud infrastructure is the *Infrastructure virtualization layer*. A number of existing approaches deal with the Edge device virtualization, exposing them to the upper layers on different levels of abstraction. Most relevant approaches for our discussion are centered around Unikernels and kernel-supported virtualization, which is discussed in Section 4.2. Other related approaches, such as software-defined networking (SDN) and semantics-based data integration are discussed in Chapter 2.

The *Middleware* is a crucial part of software-defined IoT Cloud infrastructure and, in general, its main responsibility is to provide a uniform representation of the underlying (virtual) infrastructure resources as well as to enable delivery and consumption of such resources. This layer needs to provide mechanisms and tools for infrastructure provisioning, managing configuration models and deployment of applications. The middleware relies on and utilizes a number of different components. In the following, we only briefly discuss those components since they are out of the scope of this book, although they are the main focus of numerous research and industry approaches, e.g., [54], which can be used to complement our approach.

The *Device management and orchestration* component is generally responsible for supporting discovery and management of physical Edge devices (e.g., detecting newly connected devices), monitoring their status, but also mapping and allocation of virtual resources to the underlying devices. The *Repositories* are used to provide persistent storage facilities for configuration models, infrastructure automation scripts and software-defined units, which are delivered and deployed on the devices by the middleware. The *Identity management and access control* generally deals with assigning and managing dedicated, unique names (IDs) to individual devices, but also provides security techniques to determine which devices are permitted to be provisioned as IoT Cloud resources.

4.1.1 Research Challenges

Utility-based provisioning is a well-established and proven concept in cloud computing [24, 101]. Among other things it requires: on-demand, self-service usage models; ubiquitous access to a shared pool of configurable resources, which can be customized to exactly meet application requirements; as well as autonomous and automated allocation of the consumed resources. However, given the previously described properties of the IoT Cloud, realizing these features in the context of IoT Cloud systems is a non-trivial task which creates a number of challenges that need to be addressed.

One of the main challenges is to support *the on-demand, self-service usage model*, because it requires support for uniform interactions with the large-scale, heterogeneous IoT Cloud resource pool. This could potentially be achieved by virtualizing and encapsulating the IoT Cloud resources into well-defined APIs and allowing the users to access such resources on multiple levels of abstraction. However, in this case the middleware (Figure 4.1) needs to provide support for the non-trivial task of managing such virtual resources and their APIs and mediating all communication with the heterogeneous devices.

Assuming that IoT Cloud resources are accessible in a uniform manner, another challenge is to enable the users to *automatically provision IoT Cloud resources*. However, the strong dependence of IoT Cloud applications on specific properties of the underlying devices and novel resource features intrinsically prevent consumption of IoT Cloud infrastructure as traditionally generic compute or storage utilities. This requires comprehensive provisioning support on multiple levels such as the infrastructure-, platform- and application-levels. One way to achieve this is by utilizing provisioning workflows [84] (Figure 4.1 (top)). The main advantage of the workflow approach is that it allows for nested provisioning workflows (shown as dotted nodes in the figure), which are well suited for multi-level provisioning. However, to support their execution on a large resource pool the middleware needs to enable elastically scalable execution of the provisioning tasks.

Enabling ubiquitous access to the large, geographically distributed resource pool is yet another challenge since it demands *logically centralized interaction* with underlying devices. However, since the underlying devices are inherently dispersed, the middleware needs to be distributed across the resource-constrained devices, thus and optimized for such constrained execution environments. Moreover, to support customization of such resources, the middleware needs to support *management of application components and configuration models*, but also provide suitable mechanisms to dynamically deliver and (re)enforce the configuration models inside the Edge devices.

4.2 IoT Cloud Provisioning Middleware

With respect to the components presented in Figure 4.1, the main focus of this chapter is the Middleware layer. The main purpose of our middleware is to facilitate implementing and executing provisioning workflows in IoT Cloud systems, by addressing the previously described challenges and enabling the remainder of the design principles introduced in Chapter 3. Support for multi-level provisioning is thoroughly discussed in Section 4.3. At the moment it is important to note that IoT Cloud provisioning involves two main tasks: i) *allocating and deploying Software-Defined Gateways (SDGs)*, which are a special type of the aforementioned software-defined IoT units, and ii) *customizing Software-Defined Gateways with application-specific artifacts*.

Figure 4.2 gives a high-level architecture overview of our middleware. Generally, the provisioning middleware is designed based on the microservices architecture [108] and it is distributed across the Cloud and Edge devices. The main components of the provisioning middleware include: i) the *Software-Defined Gateways*, ii) the *Provisioning and Virtual Buffers Daemons* that run in Edge devices and iii) the *Provisioning Controller* which runs in the Cloud. In the remainder of this section, we discuss these components in more detail.

4.2.1 Software-Defined Gateways

Software-defined gateways are one particular type of software-defined IoT units and their main purpose is to support virtualizing the IoT Cloud compute resources, most notably Edge devices, in order to provide isolated and managed application execution environments. Our middleware does not support building custom SDGs from

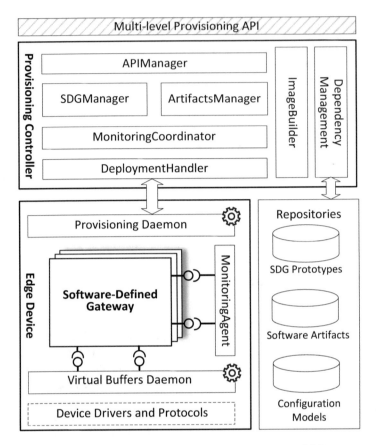

Fig. 4.2: Architecture overview of the provisioning middleware

scratch, instead it provides so-called SDG prototypes and the required mechanisms to customize them, based on application-specific requirements. At their core SDG prototypes define an isolated runtime environment for the SDGs and application-specific components. To this end, the main purpose of SDG prototypes is to provide isolated namespaces as well as to limit and isolate resource usage such as CPU and memory. Therefore, the SDG prototypes are used to bootstrap higher-level SDG functionality. In Figure 4.3 the double line shows the virtual boundaries of the SDG prototypes. It is important to mention that SDG prototypes do not propose a novel virtualiza-ton solution, but rely on proven techniques, namely kernel-supported virtualization approaches, which offer a number of lightweight execution environments/drivers such as LXCs, libvirt-sandbox or even chroot, generally referred to as containers that can be used to "wrap" SDGs. Conceptually, virtualization choices do not pose any limitations, because by utilizing the well-defined APIs, our SDGs can be dynami-cally configured, provisioned, interconnected and deployed, at runtime. The SDG prototypes are hosted in the IoT Cloud and enriched with functional and provisioning capabilities, which are exposed via well-defined APIs. A number of middleware components (Figure 4.3) are pre-installed (except for Artifact Packages) in each SDG prototype in order to support such APIs. Next, we discuss these components in more detail.

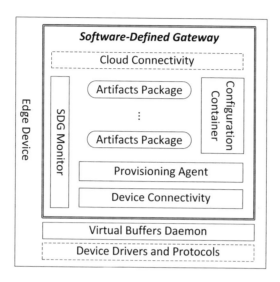

Fig. 4.3: Software-defined gateway architecture

4.2.1.1 Artifact Packages

Generally, IoT Cloud applications consist of different application components and supporting files (e.g., libraries and binaries), which we refer to as application-specific artifacts. Such artifacts are deployed, configured and executed inside software-defined gateways. Generally, our provisioning middleware does not make any assumptions about the application model or concrete artifact implementations. However, in order to enable automated artifact provisioning, it requires them to be packaged as shown in Figure 4.4. There are two important things to mention here. First, the Artifact Package needs to contain a set of provisioning directives with all the necessary instructions such as installing and uninstalling the package. When a provisioning workflow submits a provisioning request, the middleware maps the request to a concrete implementation of provisioning the directive. To support implementing such directives, previously we have introduced a lightweight provisioning DSL [122]. Second, the packages contain meta-information such as artifacts' hardware requirements and exposed APIs. The specification of the APIs is optional, but they are needed by the middleware if an application wants to completely delegate management of its configuration models to the middleware, as we discussed in Chapter 3.

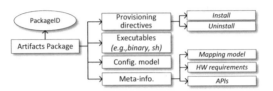

Fig. 4.4: Artifacts package structure

4.2.1.2 Provisioning Agent

All packages that are not pre-installed on the Edge devices have to be provisioned by the framework during runtime. For this purpose, our middleware provides a lightweight *Provisioning Agent*, which is pre-installed inside SDGs. The agent continuously runs in each SDG and manages local artifact packages. The main responsibility of the provisioning agent is to periodically inspect the Provisioning Controller (Figure 4.2) update queue, download the artifact packages and execute directives referenced in provisioning workflows. Additionally, the agent acts as a local interpreter of provisioning directives specified via our aforementioned provisioning DSL. The agent is also responsible for handling various requests initiated by the Provisioning Controller, by triggering the required actions in SDGs such as creating a snapshot of the current device state via the SDGMonitor and uploading the snapshot to the Controller. The SDGMonitor is discussed together with the Monitoring Agent later in this section.

4.2.1.3 Device Connectivity

The SDGs are deployed on Edge devices with limited privileges in the sense that they are not permitted to directly access the hardware. An obvious reason for such a limitation is security, but also resource contentions and customization requirements, since we can have multiple SDGs executing in the same Edge device simultaneously. To enable applications to access the underlying devices, e.g., sensors, SDG offers a *Device Connectivity* component. The main part of Device Connectivity is an SDG endpoint, which exposes the devices to the SDG and enables service-based interaction with them. The SDG endpoint is a single point of interaction with the underlying *Virtual Buffers Daemon* (Figure 4.3) and at the moment, it is defined up to the transport layer. For this reason the device connectivity component provides a pluggable connectivity layer, which is by default preconfigured with our custom, REST-like application-level protocol. In the current prototype we have also implemented CoAP and MQTT communication protocols, but the device connectivity can be easily extended by plugging in other application-level protocols such as sMAP [37].

4.2.2 Edge Device Middleware Support

In order to support management of SDGs in Edge devices, our middleware provides lightweight components that are pre-installed and continuously run inside the Edge devices. The most important components are the Virtual Buffers Daemon and Provisioning Daemon, shown in Figure 4.2 on the left-hand side.

4.2.2.1 Virtual Buffers Daemon

We have discussed how our software-defined gateways can be used for virtualizing compute resources of the Edge devices. However, since the SDGs run with reduced privileges, the middleware also needs to virtualize access to the low-level devices such as sensors and actuators. To this end it provides the Virtual Buffers Daemon (VBD). The main purpose of the VBD is to mediate communication with devices connected to a field bus (e.g., via Modbus, CAN, SOX/DASP, I^2C or IP-based) and to provide virtually exclusive access to such device. In general, the daemon acts as multiplexer of the data and control channels, thus enabling the SDGs to have their own view of and define custom configurations for such channels. For example, a software-defined gateway can configure sensor poll rates, activate a low-pass filter for an analog sensory input or configure unit and type of data instances in the stream. Figure 4.5 depicts a simplified UML diagram of the VBD's most important components. The main concept behind the VBD is the VirtualBuffers. Generally, the main goal of the virtual buffers is to provide a virtual representation of sensors and actuators. They wrap the DeviceDrivers and share a common behavior with them, inherited through the Component Interface. For example, they can be initialized, shutdown and released. Both buffers' and drivers' lifecycles are managed by the VirtualBuffersManager. The DeviceDrivers Package contains a set of driver implementations. For readability purposes, in the figure we only show the component for the I^2C protocol, but each implementation follows similar principles. It contains a set of Ports, which is a VBD internal representation of devices attached to the bus. Such Ports are dynamically instantiated by the VirtualBuffersManager at device bootup during driver initialization phase, based on the provided PortConfig. At the moment, PortConfig is specified as a JSON file that contains the metadata such as port class (e.g., analog in), name and hardware-related data, e.g., multiplexer address or value correction constants. One of the limitations of the current implementation is that it does not support dynamic device reconfiguration, meaning that if low-level configurations change the VBD must be restarted. Moreover, a virtual buffer references a set of Gatherers and can contain an optional AdapterChain. Generally, a gatherer is a higher-level representation of a port. For example, in case of a sensing device the gatherer represents the most recent value of the hardware interface. To support SDG-specific configurations such as sensor poll rate, filters or scalers, each virtual buffer can have an AdapterChain. Adapter chains reference different Adapters, which are specified and parametrized via BufferConfig. For example, a raw sensing value is passed through such an adapter chain before being delivered to

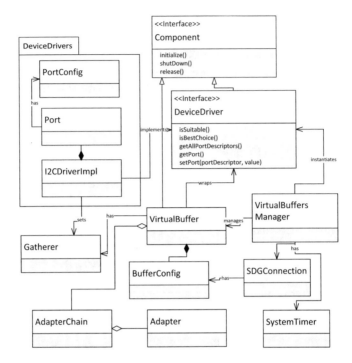

Fig. 4.5: Simplified UML diagram of the Virtual Buffers Daemon

an SDG. Finally, the VBD is responsible for instantiating and maintaining an open communication channel with software-defined gateways (via SDGConnection) and keeping track of the mappings among the SGDs and their VirtualBuffers.

4.2.2.2 Provisioning Daemon

So far, we have tacitly assumed that SDGs are readily available and deployed in Edge devices. However, this is naturally not the case, thus the SDGs need to be dynamically allocated, instantiated and deployed on Edge devices. These tasks are the shared responsibility of the Provisioning Daemon and the Provisioning Controller.

Generally, the Provisioning Daemon serves two main purposes: i) It continuously runs in each Edge device and provides functionality to remotely manage the SDGs. The remote endpoint is the middleware's Provisioning Controller. ii) It acts as a local proxy to the provisioning agents running inside each SDG, mediating all the previously described provisioning communication with SDGs (Provisioning Agents). At its core the provisioning daemon has a lightweight httpd server to allow for bidirectional communication between the Provisioning Controller and the Edge devices (i.e., SDGs). It is designed as a pluggable component, which relies on the existing support for managing shared hosting domains (i.e., containers) such as

Docker, LXD or virsh. In this context, the main components of the Provisioning Daemon are an InvocationMapper and a set of plug-in components called Connectors. Among other things, the InvocationMapper is responsible for handlin the provisioning requests from the controller and mapping them to the corresponding Connector as well as to obtain the required SDG prototypes form the Repositories and locally manage their images. The connectors act as wrappers of the underlying mechanisms for managing SDGs, exposing them to the InvocationMapper via uniform APIs. Therefore, to use a different virtualization solution for SDGs, one only needs to develop the needed connector and register it with the InvocationMapper. Second, the provisioning daemon mediates communication with the SDG provisioning agents. To support this, it manages local network interfaces of SDGs and behaves like a transparent proxy for inbound communication. Regarding outbound communication the Provisioning Daemon treats the monitoring responses in a particular manner. It intercepts the monitoring information delivered by SDGMonitors and enriches it with the current device state information, delivered by the MonitoringAgent (Figure 4.2). The *MonitoringAgent* is used to collect meta-information about the SDGs such as ID, but also to continuously monitor the underlying system via the available interfaces in order to provide dynamic device information. To this end, it executes a sequence of runtime monitoring actions to complete the dynamic device-state snapshot. For example, such actions include: currently available disk space, available RAM, firewall settings, environment information, list of processes and daemons, as well as a list of currently installed and running SDGs. The created snapshots are transmitted to the Provisioning Controller periodically or on request. The device snapshot is also used by the InvocationMapper to determine whether a new SDG can be instantiated and deployed on the Edge device, since current virtualization management solutions only provide rudimentary support in this regard.

4.2.3 Cloud-Based Provisioning Controller

The Provisioning Controller (Figure 4.2) is the cloud counterpart part of our middleware. It provides a mediation layer that enables the users to interact with the IoT Cloud in a conceptually centralized fashion, without worrying about geographical distribution and heterogeneity of the underlying Edge devices. Internally, the Provisioning Controller comprises several microservices: *APIManager, Monitoring-Coordinator, SDG- and ArtifactsManager, DeploymentHandler, ImageBuilder and DependencyManagement.*

The main responsibility of the *APIManager* is to manage the *Multi-level Provisioning API*, i.e., it encapsulates the middleware provisioning capabilities in well-defined APIs and handles all API calls from user-defined provisioning workflows. Although our middleware provides multi-level provisioning support, this distinction is only relevant to the middleware internal components, since APIManager hides all such details from the users, who effectively observe only simple API calls and corresponding responses. Therefore, the APIManager is responsible for resolving incoming requests, mapping them to the respective handlers, i.e., *SDGManager* or *ArtifactsManager* (depending on the request type), and delivering results to the calling workflow. Among other things, the actions performed by these managers involve selecting requested SDGs or artifacts by querying the corresponding SDG- and Artifacts-Repository, building the package images and deliver them to the Edge devices. In Section 4.3, we describe this process in more detail.

Since the majority of application artifacts and SDG images are not readily available in Edge devices, the *DeploymentHandler* is responsible for delivering them to the Edge devices (i.e., *Provisioning Daemons*) or SDGs (i.e., *Provisioning Agents*) at runtime. The DeploymentHandler relies on the *DependencyManagement service* to resolve the required artifact dependencies and *ImageBuilder* to prepare (package and compress) them into deployable images. Resolving the dependencies in the cloud is particularly useful, because it saves a lot of processing and networking, from the perspective of the whole IoT Cloud infrastructure, since otherwise each Edge device would have to perform the same set of actions, e.g., downloads. Furthermore, as opposed to fully fledged OS distros, Edge devices usually provide limited support in terms of packaging or updating tools, since they often run stripped-down userland such as BusyBox.

To create the aforementioned deployable images, our middleware uses the *ImageBuilder*. In order to build an image, the builder performs the following steps: (i) retrieve gateway-specific information from the IoT gateway management, (ii) use the dependency management service to gather a list of suitable plans, (iii) based on the plan, build an image, (iv) if the build was successful, hand over to the deployment handler, (v) if the build failed try the next plan in the list. Finally, all device state-snapshots are maintained by the MonitoringCoordinator, which manages static device meta-information and periodically sends monitoring requests to the *MonitoringAgent* in order to obtain runtime snapshots of current device state. The role of the MonitoringCoordinator and the MonitoringAgents is described in more detail in Section 4.3.

4.3 Runtime Mechanisms for Multi-level Provisioning in the IoT Cloud

4.3.1 Runtime Execution of Provisioning Workflows

In general, to provision (a part of) an IoT Cloud application a user might design a workflow resembling our example provisioning workflow shown in Figure 4.6 at the top. Individual actions of such a workflow usually reference specific provisioning capabilities, exposed via the middleware APIs, and rely on the middleware to support their execution. Usually, the main execution thread of provisioning workflows (denoted by the solid lines in our example provisioning workflow), represents provisioning directives for the infrastructure level, such as to deploy an SDG of a specific

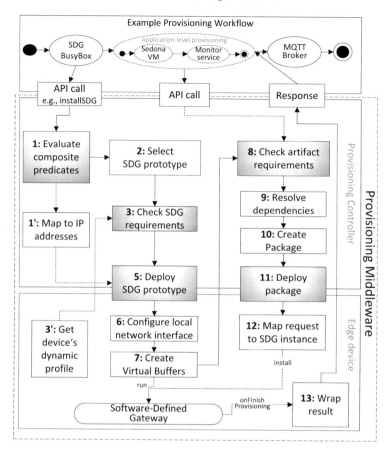

Fig. 4.6: Runtime execution of a provisioning workflow

type on Edge devices (in this case based on BusyBox) or spinnup a cloud-based Message Queue Broker, e.g., MQTT Broker. The sub-workflows (denoted by dashed lines in the same example), are mainly used to specify application-level provisioning directives. As previously mentioned, this involves customizing the SDGs with the application-specific artifacts and configuration models. For example, this can involve deploying, configuring and starting an application service.

Figure 4.6 also depicts a simplified sequence of steps performed by the middleware when executing a provisioning workflow. For the sake of clarity, we omit several steps and mainly focus on showing the most common interaction, e.g., we assume no errors or exceptions occur and we do not show interaction with the Repositories.

A provisioning workflow requests an application artifact or an SDG by specifying their respective IDs (currently consisting of a name and a version number) and a specific Edge device ID. Next, the workflow invokes a specific API, e.g., to install or uninstall the artifact. At this point the middleware attempts to execute the specified provisioning directive. The Steps 1 to 7 in Figure 4.6 depict the most important actions performed by our middleware in order to support an infrastructure-level provisioning request, e.g., to deploy, instantiate and start an SDG in an Edge device. Therefore, the middleware performs the following actions: i) The *APIManager* initially evaluates the composite predicates (described later in this section) in order to determine a set of devices on which the SGD will be deployed; ii) The *SDGManager* selects a device-compatible SDG prototype and registers it with the *DeploymentHandler*; iii) The *MonitoringCoordinator* together with the *MonitoringAgent* checks the SDG against the current device-state snapshot; iv) The *DeploymentHandler* transfers the SDG prototype image to the *Provisioning Daemon*; v) The *ProvisioningDaemon* configures the SDG's local network interface (based on the supplied mapping model), starts the SDG and registers the new SDG instance with the *Virtual Buffers Daemon*; vi) Finally the *Virtual Buffers Daemon* allocates a set of dedicated virtual buffers and creates a dedicated SDGConnection handler. At this point the SDG instance is running in the Edge device and it is performing internal initialization actions such as starting the Configuration Container, the Provisioning Agent and its local SDG Monitor. After the final initializations the SDG transmits its initial device state to the controller and it is ready to handle application-level provisioning requests.

To support an application-level provisioning request the provisioning middleware performs the following actions (steps 8 to 13 in Figure 4.6): i) Similarly to Step 3 each application artifact is checked against the current SDG-state snapshot, delivered by the *SDG Monitor*; ii) The *Dependency Management Service* resolves runtime dependencies of the artifact; iii) The *PackageManager* builds a deployable image and registers it with the *DeploymentHandler*; iv) Similarly to step 5 the *DeploymentHandler* delivers the image to the *Provisioning Daemon*; v) Finally, the *Provisioning Daemon* transparently forwards the image to the SDG's *Provisioning Agent*, which installs the package locally in the SDG. In the remainder of the section we describe the most important runtime mechanisms in more detail.

4.3.2 Evaluating Composite Predicates

While describing the main steps of the provisioning process, we have mostly focused on the steps performed for a single device and a single SDG. However, usually the provisioning workflows are meant to provision multiple devices, e.g., that share some common properties or belong to the same organization. Therefore, the same provisioning logic should be applicable regardless of specific devices. In this context, it is particularly important to support designing generic provisioning workflows, in the sense that such workflows should be defined independently of the Edge devices, e.g., without referencing device IDs. One of the main preconditions for this is to support the users in dynamically delimiting the range of provisioning actions. In our middleware this is achieved by allowing the users to specify the required device properties, as a set of composite predicates. Such predicates reference device or SDG meta-information and are used to filter out only the matching devices that meet the specified criteria. These predicates are specified by the users and delivered to the middleware in a provisioning request as POST parameters.

To bootstrap delimiting the range of a provisioning action, our middleware maintains a set of available devices for a particular user. The current prototype always considers all the connected devices, since at the moment there is only a limited support for managing the device identities and the access control. However, this is not a conceptual drawback and there are many available solutions that can be used to provide this functionality (as discussed in Section 4.1). The predicates are applied on this set, filtering out all resources that do not match the provided attribute conditions. The middleware uses the resulting set of resources to initiate the provisioning actions with *the SDGManager and the AtrifactsManager*. These managers are also responsible for providing support for gathering results delivered by the *Provisioning-Daemons* and the *ProvisioningAgents*, once the provisioning action is completed (cf. Figure 4.6 step 13). This is needed since after the resources are selected, provisioning actions are performed in parallel and the results are asynchronously delivered to provisioning workflows.

4.3.3 Artifacts and SDGs Prototypes Runtime Validation

Since we are dealing with resource-constrained devices, before deploying an SDG or application artifact the middleware needs to verify that the component can be installed on a specific device, e.g., that there is enough disk space available. This happens during step 3 (Check SDG requirements) and step 8 (Check artifact requirements). To this end, the *MonitoringCoordinator* first queries the Repositories. Besides the artifact binaries and SDG prototypes, the repositories store corresponding meta-information, such as the required CPU instruction set (e.g., ARMv5 or x86), disk space and memory requirements. After obtaining the meta-information our middleware starts building the current device-state snapshot. This is done in two stages. First, the device features catalog is queried to obtain relevant static information, such as CPU architecture, kernel version and installed userland (e.g., BusyBox [23]) or OS. Second, the *MonitoringCoordinator* in coordination with the *MonitoringAgent* and *SDGMonitor* executes a sequence of runtime profiling actions to complete the dynamic device-state snapshot. For example, the profiling actions include currently available disk space, available RAM, firewall settings, environment information, list of processes and daemons, and list of currently installed capabilities. Finally, when the dynamic device snapshot is completed, it is compared with the SDG's/artifact's meta-information in order to determine whether it is compatible with the device. In this context, the middleware performs in a similar fashion to a fail-safe iterator, in the sense that it works with snapshots of device states. For example, if something changes on the device side, during step 3 or step 8, it cannot be detected by the middleware and in this case its behavior is not defined. Since we assume that all the changes to the underlying devices are performed exclusively by our middleware, this is a reasonable design decision. Other errors, such as failure to install an artifact, in a specific SDG, are caught by the middleware and delivered as notifications to the provisioning workflow, so that they do not interrupt its execution. With this approach the middleware is capable of making autonomous decisions about the provisioned resource. This is one of the main preconditions for supporting automated execution of provisioning workflows, but also for enabling an on-demand, self-service provisioning model, since our middleware does not make any implicit assumptions such as user awareness of device properties nor does it require them to manually interact with the underlying devices.

4.3.4 Provisioning Models

One of the main goals of our middleware is to support on-demand resource consumption. Previously, we have discussed some of the key preconditions such as the ability to execute multiple SDGs inside an Edge device as well as to dynamically and automatically determine whether an SDG or an application package is suitable for a particular device, based on the monitoring device-state snapshot. In the following we discuss the provisioning models currently supported by the middleware prototype and discuss some possible optimizations. After the *MonitoringCoordinator* determines an SDG/package is compatible with Edge devices, the middleware needs to create an SDG or Artifact image and deliver it to these devices (steps 5 and 11 in Figure 4.6). This process requires the middleware to make the following decisions: what to deliver to the devices, how to deliver it and where to host the image. Therefore the image delivery process is structured along these three main phases.

4.3.4.1 Delivery Models

In the first phase, the middleware needs to choose whether to deliver a complete image or only a download script. In the first case the *ImageBuilder* creates an SDG or an Artifact image, which is essentially a compressed SDG Prototype Artifact Package. This image is then registered with the *DeploymentHandler* by a corresponding manager, which transfers the whole image to the *ProvisioningDaemon*. In the second case the process is done in a similar fashion, but in addition to the image the *ImageBuilder* also generates a download script. The main part of this script is a URL of the location where the actual image resides. Instead of the whole image only this script is sent to the *ProvisioningDaemon*, so it can download and install the image. Since both of these approaches have their advantages [138], the middleware leaves it to the user to make a decision, i.e., to select the most suitable approach and pass it as a configuration parameter in the provisioning request.

4.3.4.2 Deployment Models

In the second phase the *DeploymentHandler* deploys the image (or the download script) to the device. We support two different deployment strategies. The first strategy is pull-based, in the sense that the image is placed in a queue and remains there for a specified period of time (TTL). Both *ProvisioningDaemons* and *ProvisioningAgents* periodically inspect the queue for new provisioning requests. When a request is available, the device can pull the new image when it is ready, e.g., when the load on it is not too high. Although a provisioning workflow can specify the image priority in the queue, if a device is busy over a long period of time, e.g., there is not enough disk space to install an SDG, this can lead to request starvation, blocking the execution of the provisioning workflow. For this reason our middleware also supports push-based deployment. In this case, instead of waiting in the queue, an image is immediately pushed onto device. This gives greater control to the provisioning framework, but since the previously described image runtime validation performs in a fail-safe manner, push-based deployment can lead to undesired behavior. Therefore, when using this strategy a provisioning workflow should also provide compensation actions, to return the device to the previous state. Naturally, these two strategies can be used to create hybrid deployment strategies, such as using the pull-based approach for SDG prototypes and the push-based approach for application artifacts, because pushing artifacts is particularly useful for security updates of hot fixes in SDGs.

4.3.4.3 Placement Models

Finally, the middleware decides where to host the image. This largely depends on a specific deployment strategy, but also on the delivery model. For example, for push-based deployment the *DeploymentHandler* stores the images in-memory, also the download scripts are always kept in-memory, but in case of a pull-based strategy, images are usually hosted in middleware local *Repositories*. However, it is not difficult to imagine more complex provisioning models, which can be put in place in order to optimize the provisioning process, e.g., to save bandwidth. For example, to achieve this, our middleware could easily utilize proven technologies such as Content Delivery Networks (CDNs), Cloudlets or micro data centers. One way of accomplishing this is to deliver a download script to a set of Edge devices and push an image onto a Cloudlet, residing in the proximity (single-hop) of these devices. The *ProvisioningDaemon* could then use the pull-based approach to obtain the image.

4.4 Prototype Implementation & Evaluation

In the following experiments we show two main performance aspects of our provisioning middleware, support for: i) *scalable execution of the provisioning workflows* (hundreds of Edge devices) and at the same time ii) *middleware suitability for*

Fig. 4.7: An example of our gateways for Building Management Systems

constrained devices in terms of resource consumption, i.e., its memory and CPU usage.

4.4.0.1 Applications

In the experiments we used two real-life applications from our Building Management System, described in Chapter 3. For our experiments, it is important to note that the first application (SAPP) is written in Sedona [175] and its size is approximately 120 KB, including the SVM and the application (.sab, .sax, .scode and Kits files). The second application (JAPP) is JVM-based (compact profile2) and its size including all binaries, libraries and the JVM is around 14 MB.

Additionally, for the experiments we have developed a SDG prototype, based on BusyBox, which is a very lightweight Linux userland. The SDG prototype is specifically built for Docker's libcontainer virtualization environment and is approximately 1.4 MB in disk size (without applications).

4.4.0.2 Experiment Setup

In order to evaluate middleware performance regarding resource usage, we built 15 physical gateways (Figure 4.7) and installed them throughout our department. The gateways are based on Raspberry Pi 2, with ARMv7 CUP and 1 GB of RAM. They run Raspbian Linux 8 (based on Debian "Jessie") on Linux Kernel 4.1.

In order to evaluate how our middleware behaves in a large-scale setup, we created a virtualized IoT cloud testbed based on CoreOS [34]. In our testbed we used Docker containers to mimic physical gateways in the cloud. These containers are based on a snapshot of a real-world gateway, developed by our industry partners. For the experiments, we deployed a CoreOS cluster on our local OpenStack cloud. The cluster consists of 16 CoreOS 444.4.0 VMs (with 4 VCPUs and 7 GB of RAM), each running approximately 250 Docker containers. The Provisioning Controller and the Repositories were also deployed in our Cloud on 3 Ubuntu 14.04 VMs (with 2 VCPUs and 3 GB of RAM).

Finally, since the physical gateways are attached to our department network, in order to connect them to the cloud network (but avoid potential security risks), we created a network overlay based on Wave routers [181].

4.4.0.3 Experiments

Middleware resource consumption at the Edge. Initially, we show the performance of the most important middleware components that continuously run in edge devices, namely the *ProvisioningDaemon* and the VirtualBuffersDaemon. The *MonitoringAgent* is not considered in our experiments, since it only periodically executes to create device-state snapshots, thus it does not have statistically significant impact on the performance. We also do not discuss SDG resource consumption, since it is largely application dependent, but also depends on the underlying virtualization choices. However, it is important to mention that the runtime overhead of middleware components running in the SDGs is almost negligible, since it is less than 1 MB. The main goal of the following experiments is to demonstrate the validity of our approach

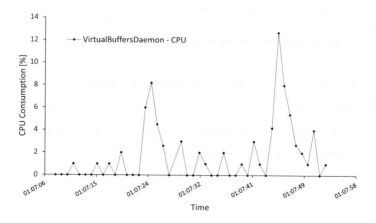

Fig. 4.8: CPU consumption of the VirtualBuffersDaemon

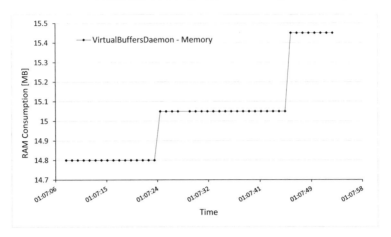

Fig. 4.9: Memory consumption of the VirtualBuffersDaemon

w.r.t. resource-constrained devices, since we do not claim that it outperforms related approaches that provide functionality that partially overlaps with our middleware.

Figure 4.8 and Figure 4.9 respectively show the CPU and memory usage of the VirtualBuffersDaemon, over a period of time. There are several important things to notice here. When there are no SDGs (applications) running in the gateway the daemon is mainly idle, i.e., it only periodically pulls the underlying drivers for device status and on average its CPU consumption is less than 2%. This can be observed in Figure 4.8, before the first peak. The two peaks represent SDG deployments for the two applications. The first peak happens when the Sedona-based application is deployed and the second peak signals the deployment of the Java-based application. Since SAPP requires fewer number of sensors than JAPP, the daemon needs to allocate and configure fewer virtual buffers, hence the difference in the two peaks. However, in both cases the maximum CPU usage of the daemon is below 14% and it lasts only a few seconds. For the same scenario we have measured the daemon's memory usage. Figure 4.9, shows the total memory of daemon's JVM process (with heap memory, Perm Size and stack). Initially, we notice that in the idle state the daemon consumes a little bit under 15 MB of RAM (the initial heap size is configured to a minimum of 1 MB), that can be considered a low memory footprint. We also observe that memory consumption behaves in a similar manner to CPU consumption. This is represented by the two distinct jumps in memory usage (Figure 4.9). The increase in memory usage is due to newly allocated virtual buffers, adapters (heap) and SDGConnections (stack). The reason for the difference is the same as above. Finally, we notice a monotonic growth of memory usage. The reason for this is that the Daemon does not trigger garbage collection, since both SDGs are running and using the buffers, however after an application is stopped the daemon releases its buffers. Therefore, the performance of the VirtualBuffersDaemon can be seen as suitable for resource-constrained devices.

Figure 4.10 and Figure 4.11 show the CPU and memory usage of the ProvisioningDaemon (and the used Connector for the underlying virtualization solution). In this case we only consider infrastructure-level provisioning requests, i.e., configuring and starting SDGs, since only this type of requests is explicitly handled by the ProvisioningDaemon. In Figure 4.10, we notice that in general our provisioning daemon utilizes the CPU resources sparingly, namely its CPU usage is mostly around 1%. This is due to the fact that most of the time the daemon idle, it only periodically checks for new requests from the Provisioning Controller and sends a hartbeat. The dramatic spikes in CPU usage happen only during the SDG deployment (we launched 4 SDGs on the gateway during the experiment), since this includes expensive network and computation operations, i.e., downloading SDG prototype, configuring it and starting it. However, the later two operations are performed by the Connectors which execute the commands and quickly terminate. Figure 4.11 shows the memory usage of the provisioning daemon for the same experiment. One can notice that during the experiment the memory usage of the provisioning daemon was always below 30 MB and more importantly shortly after an SDG is started the daemon releases the unused (Connector's) memory. Therefore, middleware Edge components in total require under 45 MB of memory and consume around 2% of CPU on average. We believe that this is reasonable resource utilization suitable for resource-constrained devices.

Scalable execution of provisioning workflows. The reason we put an emphasis on the scalability of our middleware is that it is one of the key preconditions for consistent realization of provisioning workflows across a large resource pool. For example, if the execution of provisioning workflows were to scale exponentially with the size of the resource pool, theoretically it would take infinitely long to have a consistent infrastructure baseline for the whole system, given a sufficiently large resource pool.

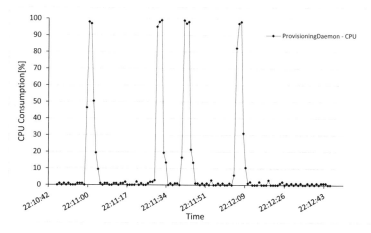

Fig. 4.10: CPU consumption of the ProvisioningDaemon

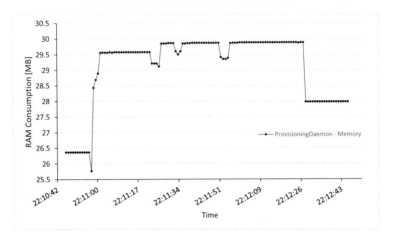

Fig. 4.11: Memory consumption of the ProvisioningDaemon

The experiment presented in Figure 4.12 and Figure 4.13 shows execution times (averaged results of 30 repetitions) of the JAPP and SAPP provisioning workflows. In the experiments we have used up to 1,000 virtual gateways for the JAPP and up to 4,000 gateways for SAPP. This corresponds to a large building management system containing dozens of big buildings (each with more than 10 floors). As a reference, the diagram also shows a plot of a trend line, which turns out to be a $nlog(n) + c$ function, extrapolated from individual experiment runs. Figure 4.12a depicts the provisioning time, i.e., execution time, of the provisioning workflow for the JAPP application. At 300 gateways we see that the initial overhead of the pushing approach is compensated and therefore the execution time decreases a little bit. From 400 to 500 gateways, the middleware reaches its maximal load. After the deployment size reaches 500, the middleware or more precisely the cloud-based controller scales out and the load balancer starts distributing the workload to the newly deployed microservices, i.e. the SDGManager, the ArtifactsManager and the DeploymentHandler. The corresponding scatter plot, depicted in Figure 4.12b, reveals that the deviations of data points are relatively small, thus on average the provisioning execution time scales almost linearly ($nlog(n)$) up to 1,000 Edge devices in this experiment.

Figure 4.13 shows the overall execution time of the SAPP provisioning workflow for different deployments (number of gateways) using the simple push-based approach. In Figure 4.13a we notice that due to the deployment scale the overall execution time got slower compared to the first experiment. This increase in the number of virtualized Edge devices, generates a lot of traffic for the underlying network infrastructure that causes slower response times and therefore the execution time of the provisioning workflow takes noticeably more time. For this scenario we changed the load-balancing strategy to allow up to 2,500 gateways before scaling out. We clearly see that up to 2,500 gateways, the execution time increases almost linearly.

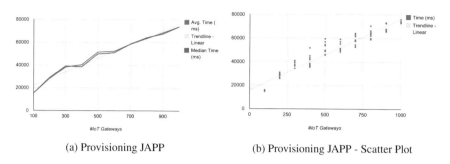

(a) Provisioning JAPP   (b) Provisioning JAPP - Scatter Plot

Fig. 4.12: Average execution time of provisioning workflow for JAPP application

When we reach 3,000 deployments, the execution time rises again, but once more starts to flatten at 4,000. When looking at the scatter plot depicted in Figure 4.13b we see that at the beginning of the experiments the deviation among data points is very small and it gets larger with an increasing number of IoT gateways. Nevertheless, we clearly see that our framework deals well with this rather large scenario and once again provides almost linear scalability.

Generally, we notice that the middleware mechanisms for workflow execution (Section 4.3.1) scale within $O(nlog(n))$ for a relatively large number of Edge devices, which can be considered a satisfactory result. We also notice that the computational overheads of the provisioning agents and daemons have no statistically significant impact on the results, since they are distributed among the underlying devices. Finally, the provisioning mechanism behaves in a similar fashion for both applications. The reason for this is that all gateways are in the same network, what can be seen as equivalent to provisioning a complex of collocated buildings.

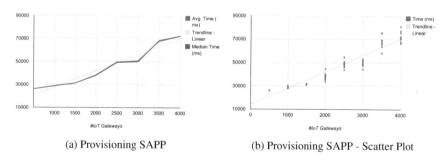

(a) Provisioning SAPP   (b) Provisioning SAPP - Scatter Plot

Fig. 4.13: Average execution time of provisioning workflow for SAPP application

4.5 Summary

In this chapter, we introduced a Smart City infrastructure provisioning middleware that enables development of generic, multi-level provisioning workflows and supports automated and scalable execution of such workflows in IoT Cloud systems. We showed how our middleware supports on-demand, self-service resource consumption by providing flexible provisioning models and support for uniform, logically centralized provisioning of Edge devices, application artifacts and their configuration models. We introduced provisioning support for software-defined gateways to enable application-specific customization of Edge devices through well-defined APIs, while preserving the benefits of proven virtualization mechanisms. The initial results of our experiments are promising, since they show that our middleware enables scalable execution of provisioning workflows across a relatively large IoT cloud resource pool and at the same time its overhead in terms of resource consumption is suitable for resource-constrained devices. Additionally, we discussed possible optimizations of the provisioning model as a direct consequence of the the the middleware's architecture. In this regard, the main advantage of middleware's architecture is reflected in its support for flexible and customizable delivery, deployment and placement models for SDGs and application artifacts. For example, it was discussed how our middleware can be configured to optimize the provisioning process by utilizing proven technologies such as CDNs.

Our middleware lays a cornerstone towards realizing our vision of utility-based provisioning of IoT Cloud systems. However, some challenges still remain to realize the utility-based provisioning paradigm. As part of our future work, we plan to address the current limitations of our middleware and the remaining challenges listed in Figure 3.1, by extending our middleware in several directions to: i) Address the mobility aspects of the Edge devices, especially focusing on dependability issues related to device mobility and mobility of software components, i.e., runtime migration of SDGs; ii) Support smarter resource allocation, i.e., optimize placement of SDGs and applications on Edge devices to include support for dynamic infrastructure properties; iii) Provide more dynamic and finer-grained resource monitoring in order to support the pay-as-you-go model; iv) Finally, we plan to extend the current prototype to address elasticity aspects for IoT Cloud systems, most notably to support elastic, on-demand scaling of the SDGs.

Chapter 5
Governing Smart City Systems

The wide and ever-growing application area of the IoT Cloud in the context of Smart Cities has led to a stronger interplay and entanglement among a variety of diverse stakeholders, with different objectives, interests and backgrounds. Various Smart City domains (sectors), such as smart building and transportation management, increasingly rely on IoT Cloud resources and capabilities to optimize their key business tasks and improve efficiency of processes and quality of life. From the application point of view IoT Cloud systems are becoming an integral enabler in optimizing urban processes, infrastructure and facilities, such as urban transportation and energy management, in order to make the cities of the future smarter and more livable. Consequently, governance issues such as security, safety, legal boundaries, compliance and data privacy concerns are being ever more strongly addressed [45, 49, 183], mainly due to their potential impact on the variety of involved Smart City stakeholders. However, such approaches are mostly intended for high-level business stakeholders, neglecting support, in terms of tools and frameworks, to realize governance strategies in large-scale, geographically distributed IoT Cloud systems. Considering the IoT Cloud from the operations management perspective, different approaches have been introduced, e.g., [189, 167, 30, 169]. For example, such approaches deal with IoT Cloud infrastructure virtualization and its management, enabling utilization of IoT Cloud computation resources and operating their storage resources. However, most of these approaches do not explicitly consider high-level governance objectives such as legal issues and compliance. This increases the risk of lost requirements or causes over-regulated systems, potentially increasing costs and limiting opportunities in future Smart Cities. Therefore, current approaches to IoT governance usually addresses the *Internet* part of the IoT, e.g., in the context of Future Internet services[1], while operations processes mostly deal with *Things* (e.g., in [33]) as additional resources that need to be operated. The governance objectives (law, compliance etc.) are not easily mapped to operations processes (e.g., querying sensory data streams or adding/removing devices), so that contemporary models, which assume that business stakeholders define governance objectives and operations

---

[1] http://ec.europa.eu/digital-agenda/en/internet-things

© Springer International Publishing AG 2017
S. Dustdar et al., *Smart Cities*,
DOI 10.1007/978-3-319-60030-7_5

managers implement and enforce them, are hardly feasible in IoT Cloud systems. In practice, bridging this gap between governance and operations management of IoT cloud systems poses a significant challenge for the involved stakeholders. What is more, even with perfectly aligned governance objectives, designing and realizing operational governance processes [166, 79], poses a significant challenge. Traditional operational governance approaches are hardly applicable for IoT Cloud systems, mainly due to the large number of involved stakeholders, novel requirements for shared resources and capabilities, dynamicity, geographical distribution and the sheer scale of such systems. Supporting tools and mechanisms for runtime operational governance of IoT Cloud systems remain largely undeveloped, thus placing much of the burden on operations managers to perform operational governance processes.

This calls for a systematic approach to govern IoT Cloud resources and applications throughout their entire lifecycle. In this chapter[2] we introduce our GovOps methodology, conceptual model and framework to effectively manage runtime governance in Smart City systems. The main aim of GovOps is to bridge the gap between high-level governance objectives (e.g., costs, legal issues or compliance) and underlying operations processes that enforce such objectives. Therefore, GovOps mostly focuses on providing conceptual and framework support for designing and executing *operational governance processes*, which represent a subset of the overall IoT Cloud governance and incorporate relevant aspects of both high-level governance strategies and underlying operations management. We present a GovOps reference model that defines required roles, concepts, and techniques, to support seamless mapping between governance and operations, and to facilitate realizing IoT Cloud governance processes in Smart Cities. We introduce rtGovOps, which is a runtime framework for dynamic operational governance of large-scale IoT Cloud systems. Its main objective is to support GovOps managers in implementing and executing operational governance processes in IoT Cloud systems, without worrying about scale, geographical distribution, dynamicity and other characteristics inherent to such systems that currently hinder operational governance in practice. The rtGovOps framework provides runtime mechanisms and enabling techniques to reduce the complexity of IoT Cloud operational governance, thus enabling the GovOps managers to perform custom operational governance processes more efficiently in large-scale IoT Cloud systems.

The remainder of this chapter is structured as follows: Section 5.1 presents our motivating scenarios. In Section 5.2, we present the GovOps methodological approach to governance and operations management in IoT Cloud systems; Section 5.3 outlines the GovOps reference model and design process for GovOps strategies; Section 5.4 outlines the main concepts and the design of the rtGovOps framework; in Section 5.5, we explain the main runtime mechanisms of rtGovOps; Section 5.6 describes the experimental results and outlines the prototype implementation; finally, Section 5.7 provides final remarks and concludes the chapter.

---

[2] The work presented in this chapter was originally introduced by Nastic et al. in [119, 125, 118].

5.1 Research Context

Let us again consider our FMS scenario, introduced in Chapter 3, from the perspective of the involved stakeholders and governance requirements. Next, we briefly discuss some of the involved stakeholders, mainly focusing on their requirements and issues related to governing FMS applications and underlying Smart City resources.

As we have previously mentioned, FMS is responsible for managing electric vehicles deployed in different Smart City environments such as golf courses. These vehicles communicate with the Cloud via 3G or Wi-Fi networks to exchange telematic and diagnostic data. In the Cloud, FMS provides different applications and services to manage this data. Examples of such services include real time vehicle status, remote diagnostics and remote control. The FMS is currently used by the following three types of stakeholders: vehicle manufacturers, distributors and golf course managers. These stakeholders have different business models. For example, when a manufacturer only leases vehicles to customers, they are interested in the status and upkeep of the complete fleet and will perform regular maintenance, as well as monitor crashes and battery health. Golf course managers are mostly interested in vehicle security to prevent misuse and ensure safety on the golf course (e.g., using geofencing features). In general, the stakeholders rely on the FMS and its services to optimize their respective business tasks. Figure 5.1 gives a high-level overview of the FMS deployment and infrastructure. We notice that FMS runs atop a non-trivial IoT Cloud infrastructure that includes a variety of IoT Cloud resources. For our discussion, the two most relevant types of IoT Cloud resources are on-board physical gateways (G) and cloud virtual gateways (VG). Most of the vehicles are equipped with on-board gateways that are capable of hosting lightweight services such as geofencing or local diagnostics services. For legacy cars that are not equipped with such gateways, a device acting as a CAN-IP bridge is used (e.g., Teltonika FM5300[3]). In this case FMS hosts virtual gateways in the cloud that execute the aforementioned services on behalf of the vehicles.

We notice that the FMS is a large-scale system that manages thousands of vehicles and relies on diverse cloud communication protocols. Further, the FMS depends on IoT Cloud resources that are geographically distributed on different golf courses around the globe. Jurisdiction over these resources can change over time, e.g., when a vehicle is handed over from the distributor to a golf course manager. In addition, these resources are usually constrained. This is why the FMS heavily relies on cloud services, e.g., for computationally intensive data processing, fault tolerance or to reliably store historical readings of vehicle data. While the cloud offers the illusion of unlimited resources, systems on such a scale as FMS can incur very high costs in practice (e.g., of computation or networking). Finally, due to the large number of involved stakeholders, the FMS needs to enable runtime customizations of infrastructure resources in order to exactly meet stakeholder requirements and allow for operation within specified compliance and legal boundaries. Therefore, the IoT Cloud resources and applications need to be managed and governed throughout

---

[3] http://gpsgate.com/devices/fm5300

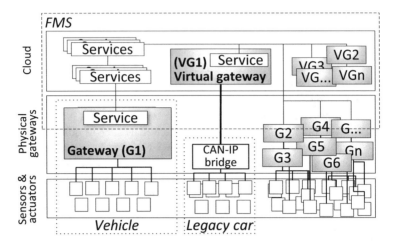

Fig. 5.1: Overview of FMS infrastructure

their entire lifecycle. In our approach, this is captured and modeled as *operational governance processes*.

5.1.0.1 Example Operational Governance Processes

Subsequently, we highlight some basic operational governance processes in FMS that are facilitated through our framework:

- Typically, the FMS polls diagnostic data from vehicles (e.g., with CoAP). However, a golf course manager could design an operational governance process that is triggered in specific situations such as in case of emergency. Such a process could, for example, increase the update rate of the vehicle sensors and change the communication protocol to MQTT in order to satisfy a high-level governance objective, e.g., the company's compliance policy to handle emergency updates in (near) real time.
- To increase fault tolerance and guarantee history preservation of vehicle data (e.g., due to governance objectives related to legal requirements), a distributor could decide to spin up additional virtual gateways in a different availability zone.
- After multiple complaints about problems with vehicles of type X, a manufacturer would need to add additional monitoring features to all vehicles of type X to perform more detailed inspections.

This is by no means a comprehensive list of operational governance processes in the FMS. However, due to dynamicity, heterogeneity, geographical distribution and the large scale of IoT Cloud systems, traditional approaches to realize even basic operational governance processes are hardly feasible in practice. This is mostly because such approaches implicitly make assumptions such as physical on-site presence, manually logging into gateways, understanding device specifics etc., which are difficult, if not impossible, to meet in IoT Cloud systems. Therefore, due to a lack of systematic approaches for operational governance in IoT Cloud systems, operations managers currently have to rely on ad hoc solutions to deal with the characteristics and complexity of IoT Cloud systems when performing operational governance processes.

5.2 GovOps – A Novel Methodology for Governance and Operations in IoT Cloud Systems

The main objective of our GovOps approach *Gov*ernance and *Op*eration*s*) is twofold. On the one side it aims to enable seamless integration of high-level governance objectives and strategies with concrete operations processes. On the other side, it enables performing operational governance processes for IoT Cloud systems in such manner that they are feasible in practice. In general, governance objectives and operations processes define and enforce system invariants that are ideally satisfied at any point in time. The objectives and states are usually associated with rules, conditions and properties that should hold during the system's runtime. In reality, due to the dynamicity and the scale of IoT Cloud systems, this is difficult if not impossible to achieve without constantly reinforcing the objectives and desired system states, as well as adapting the processes to the ever-changing requirements of the multitude of involved stakeholders.

Figure 5.2 illustrates how GovOps relates to IoT Cloud governance and operations. It depicts the main idea of GovOps to bring governance and operations closer together and bridge the gap between governance objectives and operations processes by incorporating the main aspects of both IoT Cloud governance and operations management. To this end, we define the *GovOps principles and design process* of GovOps strategies (Section 5.3) that support determining what can and needs to be governed, based on the current functionality and features of an IoT Cloud system, and that allow for such system capabilities to be aligned with the regulations and standards. Additionally, we introduce a novel role, *GovOps manager* (Section 5.2.3) responsible for guiding and managing the design of GovOps strategies, because in practice it is very difficult, risky and ultimately very costly to adhere to the traditional organizational silos

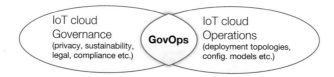

Fig. 5.2: GovOps in relation to IoT Cloud governance and operations

separating business stakeholders from operational managers. Therefore, GovOps integrates business rules and compliance constraints with operations capacities and best practices, from the early stages of designing governance strategies in order to counteract system over-regulation and lost governance requirements [49].

It is worth noting that GovOps does not attempt to define a general methodology for IoT Cloud governance. There are many approaches (Chapter 2) that define governance models and accountability frameworks for managing governance objectives and coordinating decision making processes, and these can usually be applied within GovOps without substantial modifications.

5.2.1 Governance Aspects

From our case studies, we have identified various business stakeholders such as building residents, building managers, governments, vehicle manufacturers and golf course managers. Typically, these stakeholders are interested in energy efficient and greener buildings, sustainability of building assets, legal and privacy issues regarding sensory data, compliance (e.g., regulatory or social), health of the fleet, security and safety issues related to the environments under their jurisdiction.

Depending on the concrete (sub)system and the involved stakeholders, governance objectives are realized via different governance strategies. Generally, we identify the following governance aspects: i) *environment-centric*, ii) *data-centric* and iii) *infrastructure-centric governance*.

Environment-centric governance deals with issues of overlapping jurisdictions in IoT Cloud-managed environments. For example, in our BMS, we have residents, building managers and the government that can provide governance objectives, which directly or indirectly affect an environment, e.g., a residential apartment. In this context, we need to articulate multiple governance objectives related to comfort of living, energy efficiency, safety, health and sustainability.

Data-centric governance mostly deals with implementing the governance strategies related to the privacy, quality and provenance of sensory data. Examples include addressing legal issues, compliance and user preferences w.r.t. such data.

Infrastructure-centric governance addresses issues about designing, installing and deploying IoT Cloud infrastructure. This mostly affects the early stages of introducing an IoT Cloud system and involves feasibility studies, cost analysis, and risk management. For example, it supports the decisions between introducing new hardware or virtualizing the IoT Cloud infrastructure.

5.2.2 Operations Management Aspects

Operations managers implement various processes to manage BMS and FMS at runtime. Generally, we distinguish the following operational governance aspects: i) *configuration-centric*, ii) *topology-centric* and iii) *stream-centric governance.*

Configuration-centric governance includes dynamic changes to the configuration models of the software-defined IoT Cloud systems at runtime. Example processes include: a) enabling/disabling an IoT resource or capability (e.g., start/stop a unit), b) changing an IoT capability at runtime (e.g., communication protocol), and c) configuring an IoT resource (e.g., setting sensor poll rate).

Topology-centric governance addresses structural changes that can be performed on software-defined IoT systems at runtime. For example, a) pushing processing logic from the application space towards the edge of the infrastructure; b) introducing a second gateway and an elastic load balancer to optimize resource utilization, e.g., provide more bandwidth; c) replicating a gateway, e.g., for fault-tolerance or data source history preservation.

Stream-centric governance addresses runtime operation of sensory data streams and continuous queries, e.g., to perform custom filtering, aggregation and querying of the available data streams. For example, to perform local filtering the processing logic is executed on physical gateways, while complex queries, spanning multiple data streams are usually executed on VGWs. Therefore, operations managers perform processes such as: a) placing custom filters (e.g., near the data source to reduce network traffic); b) allocating queries to virtual gateways; and c) splitting streams, i.e., sending events to multiple virtual gateways.

5.2.3 Integrating Governance Objectives with Operations Processes

The examples presented in Section 5.2.1 and Section 5.2.2 are by no means a comprehensive list of IoT Cloud governance processes. However, due to dynamicity, heterogeneity, geographical distribution and the sheer scale of the IoT Cloud, traditional approaches to realize these processes are hardly feasible in practice. This is mostly because such approaches implicitly make assumptions such as physical on-site presence, manually logging into gateways, understanding device specifics etc., which are difficult, if not impossible, to meet in IoT Cloud systems. Therefore, due to a lack of a systematic approach for operational governance in IoT systems, currently operations managers have to rely on ad hoc solutions to deal with the characteristics and complexity of IoT Cloud systems when performing governance processes.

Table 1 lists examples of governance objectives and corresponding operations management processes to enforce these objectives. The first example comes from the FMS, since many of the golf courses are situated in countries with specific data regulations, e.g., the USA or Australia. In order to enable monitoring of the whole fleet (as required by the manufacturer) the operations manager needs to understand

	Governance objectives	Operations processes
1.	Fulfill legal requirements w.r.t. sensory data in country X. Guarantee history preservation	Spin-up an aggregator gateway. Replicate VGW, e.g., across different availability zones
2.	Reduce GHG emission. User preferences regarding living comfort. Consider health regulations	Provide a configuration directives for an IoT Cloud resource (e.g., HVAC)
3.	Data quality compliance regarding location-tracking services	Choose among available services, e.g., GPS vs. GNSS (Global Navigation Satellite System) platform

Table 1: Example governance objectives and operations processes

the legal boundaries regarding data privacy. For example, in Australia, the OAIC[4] has issued a 32-page guidance as to what *"reasonable steps to protect personal information"* might include, which in practice need to be interpreted by operations managers. The second example contains potentially conflicting objectives supplied by stakeholders, e.g., building manager, end user and the government, leaving it to the operations team to solve the conflicts, at runtime. The third example hints that GNSS is usually better-suited for simultaneously working in both northern and southern high latitudes. Even for these basic processes, an operations team faces numerous difficulties, since in practice there is no one-size-fits-all solution to map governance objectives to operations processes.

Therefore, GovOps proposes a novel role, *GovOps manager*, as a dedicated stakeholder responsible for bridging the gap between governance strategies and operations processes in IoT Cloud systems. The main rationale behind introducing a GovOps manager is that in practice designing governance strategies needs to involve operations knowledge about the technical features of the system, e.g., physical location of devices, configuration and placement of queries, and component replication strategies. Reciprocally, defining systems configurations and deployment topologies should incorporate standards, compliance and legal boundaries at early stages of designing operations processes. To achieve this, the GovOps manager is positioned in the middle, in the sense that they continuously interact with both business stakeholders (to identify high-level governance issues) and the operations team (to determine operations capacities).

The main task of a GovOps manager is to determine suitable tradeoffs between satisfying the governance objectives and the system's capabilities, as well as to continuously analyze and refine how high-level objectives are articulated through operations processes. In this context, a key success factor is to ensure effective and continuous communication among the involved parties during the decision-making process, facilitating i) openness, ii) collaboration, iii) establishment of a dedicated GovOps communication channel, along with iv) early adoption of standards and regulations. This ensures that no critical governance requirements are lost and counteracts over-regulation of IoT Cloud systems.

---

[4] Office of the Australian Information Commissioner(OAIC), Australian privacy regulator.

5.2.4 Main Principles of GovOps in IoT Cloud Systems

Generally, GovOps strategies manipulate the state of IoT Cloud resources at runtime while considering governance objectives and regulations. Therefore, they can be seen as a sequence of runtime state transitions from the current state to some desired target state (e.g., that satisfies some non-functional properties, enforces compliance or exactly meets custom functional requirements). The core idea of GovOps is to provide abstractions that shield stakeholders from the complexities of the underlying system and the diversities of various legal and compliance issues, allowing them to focus on integrating governance objectives with practically feasible operations processes. To support performing such processes in IoT Cloud systems, (e.g., listed in Section 5.1), while considering system characteristics (e.g., large-scale, geographical distribution and dynamicity), GovOps relies on concepts that include:

Central point of operation (R1)   – Enable conceptually centralized interaction with the software-defined IoT Cloud system to enable a unified view of the system's operations and governance capabilities (available at runtime), without worrying about low-level infrastructure details.

Automation (R2)   – Allow for dynamic, on-demand governance of software-defined IoT cloud systems on a large scale and enable governance processes to be easily repeatable, i.e., enforced across the IoT Cloud, without manually logging into individual gateways.

Fine-grained control (R3)   – Expose the control functionality of IoT Cloud resources at fine granularity to allow for precise definition of governance processes (to exactly meet requirements) and flexible customization of IoT Cloud system governance capabilities.

Late-bound directives (R4)   – Support declarative directives that are bound later during runtime in order to allow for designing generic and flexible operational governance processes.

IoT Cloud resources autonomy (R5)   – Provide a higher degree of autonomy to IoT Cloud resources to reduce communication overhead, increase availability (e.g., in case of network partitions), enable local exception and fault handling, support protocol-independent interaction, and increase system scalability.

5.3 A Reference Model for GovOps Methodology

5.3.1 Overview of the GovOps Model for IoT Cloud Systems

To realize the GovOps approach we need suitable abstractions to describe IoT Cloud resources that allow IoT Cloud infrastructure to be (re)defined after it has been deployed. We show in Chapter 3 how this can be done with *software-defined IoT units*. The GovOps model (Figure 5.3) builds on this premise and extends our previous work with fundamental aspects of operational governance processes: i) describing

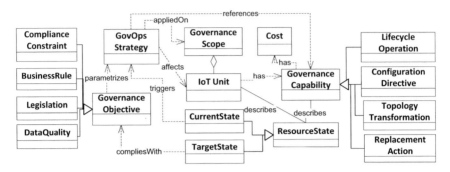

Fig. 5.3: Simplified UML diagram of GovOps model for IoT Cloud governance

states of deployed IoT resources, ii) providing capabilities to manipulate these states at runtime, and iii) defining governance scopes.

Within our model, the main building blocks of GovOpsStrategies are *Governance-Capabilities*. They represent operations which can be applied on IoT Cloud resources, e.g., query current version of software, change communication protocol and spin up a virtual gateway. These operations manipulate IoT Cloud resources in order to put an IoT Cloud system into a specific (target) state. Governance capabilities are described via well-defined software-defined APIs and they can be dynamically added to the system, e.g., to a VGW. From the technical perspective, they behave like add-ons, in the sense that they extend resources with additional operational functionality. Generally, by adopting the notion of governance capabilities, we allow for processes to be automated to a great extent, but also give a degree of autonomy to IoT Cloud resources.

Since the meaning of a resource state is highly task-specific, we do not impose many constraints to define it. Generally, any useful information about an IoT Cloud resource is considered to describe the *ResourceState*, e.g., a configuration model or monitoring data such as CPU load. Technically, there are many frameworks (e.g., Ganglia or Nagios) that can be used to (partly) describe resource states. Also configuration management solutions, such as OpsCode Chef[5], can be used to maintain and inspect configuration states. Finally, design best practices and reference architectures (e.g., AWS Reference Architectures[6]) provide a higher-level description of the desired target states of an IoT Cloud system.

The *GovernanceScope* is an abstract resource, which represents a group of IoT Cloud resources (e.g., gateways) that share some common properties. Therefore, our governance scopes are used to dynamically delimit IoT Cloud resources on which a GovernanceCapability will have an effect. This enables the governance strategies to be written in a scalable manner, since the IoT Cloud resources are not individually addressed. It also allows for backwards compatible GovOps strategies, which do not directly depend on the current resource capabilities. This means that we can move a

---

[5] http://opscode.com/chef

[6] http://aws.amazon.com/architecture/

part of the problem, e.g., faults and exceptions handling, inside the governance scope. For example, if a gateway loses a capability the scope simply will not invoke it, i.e., the strategy will not fail.

5.3.2 Design Process of GovOps Strategies

As described in Section 5.2, the GovOps manager is responsible for overseeing and for guiding the GovOps design process and to design concrete GovOps strategies. The design process is structured along three main phases: i) identifying governance objectives and capabilities, ii) formalizing strategy, and iii) executing strategy.

Generally, the initial phase of the design process involves eliciting and formalizing governance objectives and constraints, as well as identifying required *fine-grained* governance capabilities to realize the governance strategy in the underlying IoT Cloud system. GovOps does not make any assumptions or impose constraints on formalizing governance objectives. To support specifying governance objectives the GovOps manager can utilize various governance models and frameworks, such as 3P [151] or COBIT [67]. However, it requires tight integration of the GovOps manager into the design process and encourages collaboration among the involved stakeholders to clearly determine risks and tradeoffs, in terms of what should and can be governed in the IoT Cloud system, e.g., which capabilities are required to balance building emission regulations and residents' temperature preferences. To this end, the GovOps manager gathers available governance capabilities in collaboration with the operations team, identifies missing capabilities, and determines whether further action is necessary. Generally, governance capabilities are exposed via well-defined APIs. They can be built-in capabilities exposed by IoT units (e.g., start/stop), obtained from third-parties (e.g., from public repositories or in a market-like fashion) or developed in-house to exactly reflect custom governance objectives. By promoting *collaboration* and early integration of governance objectives with operations capabilities, GovOps reduces the risks of lost requirements and over-regulated systems.

After the required governance capabilities and relevant governance aspects have been identified, the GovOps manager relies on the aforementioned concepts and abstractions (Section 5.3.1) to formally define the GovOps strategy and articulate the artifacts defined in the first phase of the design process. Governance capabilities are the main building blocks of GovOps strategies. They are directly referenced in GovOps strategies to specify the concrete steps which need to be enforced on the underlying IoT Cloud resources, e.g., defining a desired communication protocol or disabling a data stream for a specific region. Also in this context, the GovOps reference model does not make assumptions about the implementation of governance strategies, e.g., they can be realized as business processes, policies, applications, or domain-specific language. Individual steps, defined in the generic strategy, invoke governance capabilities that put the IoT Cloud resources into the desired target state, e.g., which satisfies a set of properties. Subsequently, the generic GovOps strategy needs to be parametrized, based on the concrete constraints and rules defined by the governance objectives. Depending on the strategy implementation these can be realized as process parameters, language constraints (e.g., Object Constraint Language), or application configuration directives. By formalizing the governance strategy, GovOps enables reusability of strategies, promotes consistent implementation of established standards and best practices, and ensures operation within the system's regulatory framework.

The last phase involves identifying the system resources, i.e., the governance scopes that will be affected by the GovOps strategy and executing the strategy in the IoT Cloud system. It is worth mentioning that the scopes are not directly referenced in the GovOps strategies, as the GovOps manager applies the strategies on the resource scopes instead of the actual resources. Introducing scopes at the strategy level shields the operations team from directly referencing IoT Cloud resources, hence enabling designing *declarative, late-bound strategies* in a scalable manner. Furthermore, additional capabilities identified in the previous phase will be acquired and/or provisioned at this point in the underlying IoT Cloud system, whereas unused capabilities will be decommissioned in order to optimize resource consumption.

5.4 rtGovOps – A Runtime Framework for GovOps in Large-Scale IoT Cloud Systems

The main aim of our rtGovOps (*runtime GovOps*) framework is to facilitate operational governance processes for software-defined IoT Cloud systems. To this end, rtGovOps provides a set of runtime mechanisms and does most of the "heavy lifting" to support operations managers in implementing and executing operational governance processes in large-scale software-defined IoT Cloud systems, without worrying about scale, geographical distribution, dynamicity, and other characteristics inherent to such systems that currently hinder operational governance in practice. In order to facilitate the operational governance processes, while considering the characteristics of software-defined IoT Cloud systems, the rtGovOps framework follows the set of design principles, introduced in Section 5.2.4. They represent the main requirements, that need to be supported and enforced by our rtGovOps framework.

Figure 5.4 gives a high-level architecture and deployment overview of the rtGovOps framework. Generally, the rtGovOps framework is distributed across the cloud and IoT devices. It is designed based on the microservices architecture[7], which among other things enables flexible, evolvable and fault-tolerant system design, while allowing for flexible management and scaling of individual components. The main components of rtGovOps include: i) the *governance capabilities*, ii) the *governance controller* that runs in the cloud, and iii) the *rtGovOps agents* that run in IoT devices. In the remainder of this section, we will discuss these components in more detail.

---

[7] http://martinfowler.com/articles/microservices.html

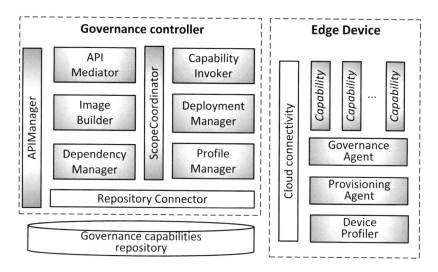

Fig. 5.4: Overview of rtGovOps architecture and deployment

5.4.1 Operational Governance Capabilities

As we described in Section 5.1, operational governance processes govern software-defined IoT units throughout their entire lifecycle. Generally, *Governance capabilities* represent the main building blocks of operational governance processes and they are usually executed in IoT devices. The governance capabilities encapsulate governance operations which can be applied on deployed IoT units, e.g., to query the current version of a service, change a communication protocol, or spin up a virtual gateway. Such capabilities are described via well-defined APIs and are usually provided by domain experts who develop the IoT units. The rtGovOps framework enables such capabilities to be dynamically added to the system (e.g., to gateways), and supports managing their APIs. From a technical perspective, they behave like add-ons, in the sense that they extend the resources with additional operational functionality. Internally, IoT devices host rtGovOps agents that behave like an add-on manager, responsible for installing/enabling, starting/stopping a capability, and managing the APIs they expose. Generally, rtGovOps does not make any assumptions about concrete capability implementations. However, it requires them to be packaged as shown in Figure 5.5. Subsequently, we highlight relevant examples of governance capabilities related to our FMS application.

- *Configuration-specific capabilities* include changes to the configuration models of software-defined IoT Cloud systems at runtime. For example: setting sensor poll rate, changing communication protocol for cloud connectivity, configuring data point unit and type (e.g., temperature in Kelvin as unsigned 10-bit integer), mapping a sensor or CAN bus unit to a device's virtual pin, or activating a low-pass filter for an analog sensory input.

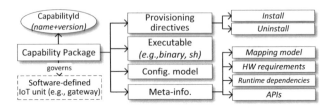

Fig. 5.5: Overview of capability package structure

- *Topology-specific capabilities* address structural changes that can be performed on the deployment topologies of software-defined IoT systems. Examples include replicating a virtual gateway to increase fault tolerance or data source history preservation and pushing data-processing logic from the application space towards the edge of the infrastructure.
- *Stream-specific capabilities* deal with managing the runtime operation of sensory data streams and continuous Complex Event Processing (CEP) queries. Therefore, to enable features such as scaling out or stream replaying, operations managers need capabilities such as: filter placement near the data source to reduce network traffic, allocation of queries to gateways, and stream splitting, i.e., sending events to multiple virtual gateways.
- *Monitoring-specific capabilities* deal with adding a general monitoring metric, e.g., CPU load, or providing an implementation of a custom metric to IoT Cloud resources.

For the sake of simplicity, we assume that the capabilities are readily available[8]. In reality, they can be obtained from a central repository, provided by a third party in a market-like fashion, or custom developed in-house.

As mentioned above, governance capabilities are dynamically added to the IoT Cloud resources. There are several reasons why such behavior is advantageous for operations managers and software-defined IoT Cloud systems. For example, as we usually deal with constrained resources, static provisioning of such resources with all available functionality is rarely possible (e.g., factory defaults rarely contain the desired configuration for FMS vehicle gateways). Further, as we have seen in Section 5.1, jurisdiction over resources (in this case FMS vehicles) can change during runtime, e.g., when a vehicle is handed over to a golf course manager. In such cases, because the governing stakeholder changes, it is natural to assume that the requirements regarding operational governance will also change, thus requiring additional or different governance capabilities. Instead of updating the whole device image at once, we reduce the communication overhead, but also enable device functionality to be changed without interrupting the system, e.g., to reboot. This provides greater flexibility and enables on-demand governance tasks (e.g., by temporarily adding a capability), which are often useful in systems with a high degree of dynamicity. Fi-

---

[8] We provide example governance capabilities under https://github.com/tuwiendsg/GovOps/

nally, executing capabilities in the IoT devices improves scalability of the operational governance processes and enables better resource utilization.

5.4.2 Operational Governance Processes and Governance Scopes

Operational governance processes represent a subset of the general IoT Cloud governance and deal with operating and governing IoT Cloud resources at runtime. Such processes are usually designed by operations managers in coordination with business stakeholders [119]. The main purpose of such processes is to support high-level governance objectives such as compliance and legal concerns, which influence the system's runtime behavior. To be able to dynamically govern IoT Cloud resources, the operational governance processes rely on the governance capabilities. This means that individual steps of such processes usually invoke governance capabilities in order to enforce the behavior of IoT Cloud resources in such a manner that they comply with the governance objectives. In this context, our rtGovOps framework provides runtime mechanisms to enable execution of these operational governance processes.

As described in Chapter 3, we use software-defined IoT units to describe IoT Cloud resources. However, these units are not specifically tailored for describing non-functional properties and available meta-information about IoT Cloud resources, e.g., location of a vehicle (gateway) or its specific type and model. For this purpose, rtGovOps provides governance scopes. The governance scope is an abstract resource that represents a group of IoT Cloud resources that share some common properties. For example, an operations manager can specify a governance scope to include all the vehicles of type X. The *ScopeCoordinator* (Figure 5.4) provides mechanisms to define and manage the governance scopes. The rtGovOps framework relies on the *ScopeCoordinator* to determine which IoT Cloud resources need to be affected by an operational governance process. Generally, the governance scopes enable implementing the operational governance processes in a scalable and generic manner, since the IoT Cloud resources do not have to be individually referenced within such a process.

5.4.3 Governance Controller and rtGovOps Agents

The *Governance controller* (Figure 5.4) represents a central point of interaction with all available governance capabilities. It provides a mediation layer that enables operations managers to interact with IoT Cloud systems in a conceptually centralized fashion, without worrying about the geographical distribution of the underlying system. Internally, the governance controller comprises several microservices, among which the most important include: *DeploymentManager* and *ProfileManager*, which are used to support dynamical provisioning of the governance capabilities, as well as *APIManager* and the previously mentioned *ScopeCoordinator* that support operational governance processes to communicate with the underlying capabilities. The *APIManager* exposes governance capabilities to operational governance processes via well-defined APIs and handles all API calls from such processes. It is responsible for resolving incoming requests, mapping them to respective governance capabilities in the IoT devices and delivering results to the calling process. Among other things, this involves discovering capabilities by querying the capabilities repository, and parameterizing capabilities via input arguments or configuration directives.

Since governance capabilities are usually not "pre-installed" in IoT devices, the *DeploymentManager* is responsible for injecting capabilities into such devices (e.g., gateways) at runtime. To this end it exposes REST APIs, which are used by the devices to periodically check for updates, as well as by the operational governance processes to push capabilities into the devices. Finally, the *ProfileManager* is responsible for dynamically building and managing device profiles. This involves managing static device meta-information and periodically performing profiling actions in order to obtain runtime snapshots of current device states.

Another essential part of the rtGovOps framework is the *rtGovOps agents*. They include: *ProvisioningAgent*, *GovernanceAgent* and *DeviceProfiler*. These agents are very lightweight components that run in all IoT Cloud resources that are managed by rtGovOps, such as the FMS vehicles. Figure 5.6 shows a high-level overview of the *GovernanceAgent* architecture. It is responsible for managing local governance capabilities, wrapping them in well-defined APIs and exposing them to the *Governance controller*. The rtGovOps agents offer advantages in terms of general scalability of the system and provide a degree of autonomy to the IoT Cloud resources.

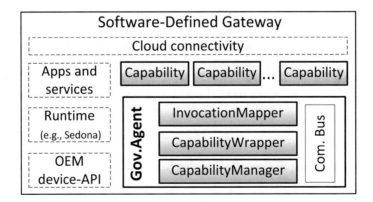

Fig. 5.6: Overview of the governance agent architecture

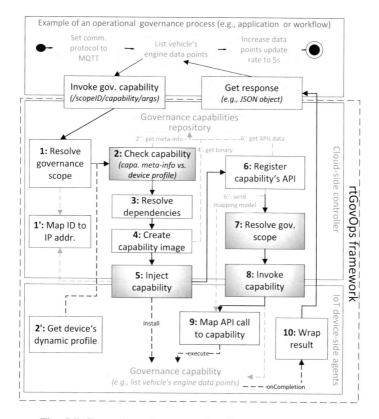

Fig. 5.7: Execution of an operational governance process

5.5 Main Runtime Mechanism of the rtGovOps Framework

Generally, the rtGovOps framework supports operations managers' handling two main tasks. First, the rtGovOps framework enables dynamic, on-demand *provisioning of governance capabilities*. For example, it allows for dynamic injection of capabilities into IoT Cloud resources, and coordination of the dynamic profiles of these resources at runtime. Second, our framework allows for runtime management of governance capabilities throughout their entire lifecycle, which among other things, includes *remote capability invocation and managing dynamic APIs* exposed to users.

As we have mentioned earlier, in order to achieve a high-level governance objective such as to enforce (part of) compliance policies for handling emergency situations, an operations manager could design an operational governance process similar to the one shown in Figure 5.7 (top). Individual actions of such processes usually reference specific governance capabilities and rely on rtGovOps to support their execution. Figure 5.7 depicts a simplified sequence of steps executed by the rtGovOps framework when a governance capability gets invoked by an operational governance process. For the sake of clarity, we omit several steps performed by the framework and mainly focus on showing the most common interaction, i.e., we assume no errors or exceptions occur. We will discuss the most important steps performed by rtGovOps below. Note that all of these steps are performed transparently to operations managers and operational governance processes. The only thing that such processes observe is a simple API call (similar to REST service invocation) and a response (e.g., a JSON array in this case). Naturally, the process is responsible for providing arguments and/or configuration directives that are used by rtGovOps to parametrize the underlying capabilities.

5.5.1 Automated Provisioning of Governance Capabilities

In order to enable dynamic, on-demand provisioning of governance capabilities, whenever a new capability is requested (i.e., referenced in an operational governance process) the rtGovOps framework needs to perform the following steps: i) the *ScopeCoordinator* resolves the governance scope to get a set of devices to which the capability will be added; ii) the *ProfileManager* checks whether the governance capability is available and compatible with the device; iii) the *Dependency Manager* resolves runtime dependencies of the capability; iv) the *ImageBuilder* creates a capability image; v) finally, the *DeploymentManager* injects the capability into devices. An overview of this process is also shown in steps $1 - 5$ in Figure 5.7.

Algorithm 5.1 shows the capability-provisioning process in more detail. An operational governance process requests a capability by supplying a capability ID (currently consisting of capability name and version) and an operational governance scope (more detail in Section 5.5.2). After that rtGovOps tries to add the capability (together with its runtime dependencies) to a device. If successful, it continues along the steps shown in Figure 5.7. The algorithm performs in a similar fashion to a fail-safe iterator, in the sense that it works with snapshots of device states. For example, if something changes on the device side inside *checkComponent* (Algorithm 5.1, lines $2 - 5$) it cannot be detected by rtGovOps and in this case the behavior of rtGovOps is not defined. Since we assume that all the changes to the underlying devices are performed exclusively by our framework, this is a reasonable design decision. Other errors, such as failure to install a capability on a specific device, are caught by rtGovOps and delivered as notifications to the operational governance process, so that they do not interrupt its execution.

5.5.1.1 Capability Checking

From the steps presented in Algorithm 5.1 *checkComponent* (lines $1 - 6$) and *injectCapability* (lines $15 - 17$) are the most interesting. The framework invokes *checkComponent* for each governance capability and all of its dependencies for the currently considered device. At this point rtGovOps verifies that the component can be installed on this specific device. To this end, the *ProfileManager* first queries the *central capabilities repository*. Besides the capability binaries, the repository stores capability meta-information, such as required CPU instruction set (e.g., ARMv5 or x86), disk space and memory requirements, as well as installation and decommissioning directives. After obtaining the capability meta-information the framework starts building the current device profile. This is done in two stages. First, the gateway features catalog is queried to obtain relevant static information, such as CPU architecture, kernel version and installed userland (e.g., BusyBox[9]) or OS. Second, the *ProfileManager* in coordination with *DeviceProfiler* executes a sequence of runtime profiling actions to complete the dynamic device profile. For example, the profiling actions include currently available disk space, available RAM, firewall settings, environment information, list of processes and daemons, and list of currently installed capabilities. Finally, when the dynamic device profile is completed, it is compared with the capability's meta-information in order to determine whetherthe capability is compatible with the device.

Algorithm 5.1: Governance capability provisioning

input : *capaID* : A capability ID.
       gscope : Operational governance scope.
result : Capability added to device or error occurred.

```
1  func checkComponent (component, device)
2  │   capaMeta ← queryCapaRepo(component)
3  │   devProfile ← getDeviceProfile(device)
4  │   status ← isCompatible(capaMeta, devProfile)
5  │   return status
6  end
   /* Begin main loop.                                          */
7  components ← resolveDependencies(capaID)
8  components ← add(capaID)
9  for device in resolveGovScope(gscope) do
10 │   for component in components do
11 │   │   if not checkComponent(component, device) then
12 │   │   │   error
13 │   │   end
14 │   end
   /* Inject capability.                                        */
15 │   capaImg ← createImg(components)
16 │   deployCapa(capaImg, device)
17 │   installCapa(capaImg) // On device-side
18 end
```

---

[9] http://busybox.net

5.5.1.2 Capability Injection

The rtGovOps *capability injection mechanism* deals with uploading and installing capabilities on devices, as well as managing custom configuration models. This process is structured along three main phases: creating a capability image, deploying the capability image on a device and installing the capability locally on the device.

1. After the *ProfileManager* determines a capability is compatible with the gateway, the *ImageBuilder* creates a capability image. The capability image is rtGovOps' internal representation of the capability package (see Figure 5.5). In essence it is a compressed capability package containing component binaries and a dynamically created runlist. The runlist is an ordered list of components that need to be installed. It is created by the *DependencyManager* and its individual steps reference component installation or decommissioning directives that are obtained from the *capabilities repository*.

2. In the second phase, *DeploymentManager* deploys the image to the device. We support two different deployment strategies. The first strategy is *pull-based*, in the sense that the image is placed in the update queue and remains there for a specified period of time (TTL). The *ProvisioningAgent* periodically inspects the queue for new updates. When an update is available, the device can poll the new image when it is ready, e.g., when the load on it is not too high. A governance process can have more control over the pull-based deployment by specifying a capability's priority in the update queue. Finally, on successful update the *DeploymentManager* removes the update from the queue. The second deployment strategy allows governance capabilities to be asynchronously *pushed* to gateways. Since the capability is forced onto the gateway, it should be used cautiously and for urgent updates only, such as increasing a sensor poll-rate in emergency situations. Finally, independent of the deployment strategy, the framework performs a sequence of checks to ensure that an update was performed correctly (e.g., compares checksums) and moves to the next phase.

3. In the final phase, the *ProvisioningAgent* performs a local installation of the capability binaries and its runtime dependencies, and performs any custom configurations. Initially, the *ProvisioningAgent* unpacks the previously obtained capability image and verifies that the capability can be installed based on the current device profile. In case the conditions are not satisfied, e.g., due to disk space limitations, the process is aborted and an error is sent to the *Deployment-Manager*. Otherwise, the *ProvisioningAgent* reads the runlist and performs all required installation or decommissioning steps.

A limitation of the current rtGovOps prototype is that it only provides rudimentary support for specifying installation and decommissioning directives. Therefore, capability providers need to specify checks, e.g., whether a configuration file already exists, as part of the installation directives. In the future we plan to provide a dedicated provisioning DSL to support common directives and interactions.

5.5.2 rtGovOps APIs and Invocation of Governance Capabilities

When a new governance capability is injected into a gateway, the rtGovOps framework performs the following steps: i) register the capability with the *APIManager*; ii) the *ScopeCoordinator* resolves the governance scope; iii) the *APIMediator* provides a mapping model to the *GovernanceAgent*; iv) the *GovernanceAgent* wraps the capability into a well-defined API, dynamically exposing it to the outside world; v) the *CapabilityInvoker* invokes the capability and delivers the result to the invoking operational governance process when the capability execution completes. A simplified version of this process is also shown in steps 6 − 10 in Figure 5.7.

Before we dive into the technical details of this process, it is worth mentioning that currently in the capabilities repository, besides the aforementioned capability meta-information and binaries, we also maintain well-defined capability API descriptions, e.g., functional, meta and lifecycle APIs. These APIs are available to operations managers as soon as a capability is added to the repository and independent of whether the capability is installed on any device. Additionally, we provide a general rtGovOps API that is used to allow for more control over the system and its capabilities. It includes the *CapabilityManager* API (e.g., list capabilities, check whether capability installed/active), capability lifecycle API (e.g., start, stop or remove capability), and the *ProvisioningAgent* API (e.g., install new capability). Listing 5.1 shows some examples of such APIs as REST-like services (version numbers are omitted for clarity).

```
1   /* General case of capability invocation.      */
2   /govScope/{capabilityId}/{methodName}/{arguments}?
3   arg1={first-argument}&arg2={second-argument}&...

4   /* Data points capability invocation example. */
5   /deviceId/DPcapa/setPollRate/arguments?rate=5s
6   /deviceId/DPcapa/list

7   /* Capabilities manager examples.              */
8   /deviceId/cManager/capabilities/list
9   /deviceId/cManager/{capabilityId}/stop
```

Listing 5.1: Examples of capabilities and rtGovOps APIs

5.5.2.1 Single Invocation of Governance Capabilities

In the following we mainly focus on explaining the steps that are performed by the rtGovOps framework when a capability is invoked on a single device. The more general case involving multiple devices and using operational governance scopes is discussed in the next section.

When a capability gets invoked by an operational governance process for the first time, the *APIManager* does not know anything about it. Therefore, it first needs to check, based on the API call (e.g., see Listing 5.1), whether the capability exists in the central capabilities repository. After the capability is found and provisioned (Section 5.5.1), the rtGovOps framework tries to invoke the capability. This involves the following steps: registering the capability, mapping the API call, executing the capability, and returning the result.

1. First, the *APIManager* registers the API call with the corresponding capability. This involves querying the capability repository to obtain its meta-information (such as expected arguments), as well as building a dynamic mapping model. Among other things, the mapping model contains the capability ID, a reference to a runtime environment (e.g., Linux shell), a sequence of input parameters, the result type and further configuration directives. The *APIMediator* forwards the model to the device (i.e., *GovernanceAgent*) and caches this information for subsequent invocations. During future interactions, the rtGovOps framework acts as a transparent proxy, since subsequent steps are handled by the underlying devices.

2. In the next step, rtGovOps needs to perform a mapping between the API call and the underlying capability. Currently, there are two different ways to do this. By default, rtGovOps assumes that capabilities follow the traditional Unix interaction model, i.e., that all arguments and configurations (e.g., flags) are provided via the standard input stream (stdin) and output is produced to standard output (stdout) or standard error (stderr) streams. This means, if not specified otherwise in the mapping model, the framework will try to invoke the capability by its ID and will forward the provided arguments to its stdin. For capabilities that require custom invocation, e.g., property files, policies or specific environment settings, the framework requires a custom mapping model. This model is used in the subsequent steps to correctly perform the API call.

3. Finally, the *CapabilityInvoker* in coordination with the *GovernanceAgent* invokes the governance capability. As soon as the capability completes, the *GovernanceAgent* collects and wraps the result. Currently, the framework provides means to wrap results as JSON objects for standard data types and it relies on the mapping model to determine the appropriate return type. However, this can be easily extended to support more generic behavior, e.g., by using Google Protocol Buffers[10].

5.5.2.2 Operational Governance Scopes

When an operational governance process gets invoked on a governance scope, the aforementioned invocation process remains the same, with the only difference that rtGovOps performs all steps on a complete governance scope in parallel instead of on an individual device. To this end, the *ScopeCoordinator* enables dynamic resolution of the governance scopes.

There are several ways in which a governance scope can be defined. For example, an operations manager can manually assign a set of resources to a scope, such as all

---

[10] http://code.google.com/p/protobuf/

vehicles belonging to a golf course, or they can be dynamically determined depending on runtime features by querying governance capabilities to obtain dynamic properties such as the current configuration model. To bootstrap defining the governance scopes, the *ScopeCoordinator* defines a global governance scope that is usually associated with all the IoT Cloud resources administered by a stakeholder at the given time. Governance scope specifications are implemented as composite predicates referencing device meta-information and profile attributes, The predicates are applied to the global scope, filtering out all resources that do not match the provided attribute conditions. The *ScopeCoordinator* uses the resulting set of resources to initiate capability invocation with the *CapabilityInvoker*. The *ScopeCoordinator* is also responsible for providing support for gathering results delivered by the invoked capabilities. This is needed since the scopes are resolved in parallel and the results are asynchronously delivered by the IoT devices.

5.6 Prototype Implementation & Evaluation

5.6.1 Prototype Implementation

In the current prototype, the rtGovOps *Governance controller* microservices are implemented in the Java and the Scala programming languages. The rtGovOps agents are based on a lightweight httpd server and are implemented as Linux shell scripts. The complete source code and supplementary materials providing more details about the current rtGovOps implementation are publicly available in Git Hub[11].

5.6.2 Setup of the Experiments

In order to evaluate how our rtGovOps framework behaves in a large-scale setup (hundreds of gateways), we created a virtualized IoT Cloud testbed based on CoreOS[12]. In our testbed we use Docker containers to virtualize and mimic physical gateways in the cloud. These containers are based on a snapshot of a real-world gateway, developed by our industry partners. The Docker base image is publicly available in Docker Hub under dsgtuwien/govops-box[13].

For the subsequent experiments we deployed a CoreOS cluster on our local OpenStack cloud. The cluster consists of 4 CoreOS 444.4.0 VMs (with 4 VCPUs and 7 GB of RAM), each running approximately 200 Docker containers. Our rtGovOps agents are preinstalled in the containers. The rtGovOps Governance controller and

---

[11] http://github.com/tuwiendsg/GovOps

[12] http://coreos.com/

[13] https://registry.hub.docker.com/u/dsgtuwien/govops-box/

capabilities repository are deployed on 3 Ubuntu 14.04 VMs (with 2VCPUs and 3 GB of RAM). The operational governance processes are executing on a local machine (with Intel Core i7 and 8 GB of RAM).

5.6.3 Governing FMS at Runtime

We first show how our rtGovOps framework is used to support operational governance processes on a real-world FMS application for monitoring vehicles (e.g., location and engine status) on a golf course (see Section 5.1). The application consists of several services. On the one side, there is a lightweight service running in the vehicle gateways that interfaces with vehicle sensors via the CAN protocol, and feeds sensory data to the cloud. On the cloud-side of the FMS application, there are several services that, among other things, perform analytics on the sensory data and offer data visualization support. In our example implementation of this FMS application, the gateway service is implemented as a software-defined IoT unit that among other things provides an API and mechanisms to dynamically change the cloud communication protocol without stopping the service.

The FMS application polls diagnostic data from vehicles with CoAP. However, in case of an emergency, a golf course manager needs to increase the update rate and switch to MQTT in order to handle emergency updates in (near) real time. This can be easily specified with an operational governance process that contains the following steps: change communication protocol to MQTT, list vehicle engine and location data points, and set data points update rate, e.g., to 5 seconds. These steps are also depicted in Figure 5.7 (top). The golf course manager relies on rtGovOps governance capabilities to realize individual process steps and rtGovOps mechanisms (Section 5.5) to execute the operational governance process.

Figure 5.8 shows the bandwidth consumption of the FMS application, which monitors 50 vehicles over a period of time. We notice two distinct operation modes: normal operation and operation in case of an emergency (emergency operation). Most notable are the two transitions: first, from normal to emergency operation and second, returning from emergency to normal operation. These transitions are described with the aforementioned operational governance process that is executed by the rtGovOps framework. The significant increase in bandwidth consumption happens during the execution of the operational governance process, because it changes the communication protocol from polling the vehicles approximately every minute with CoAP, to pushing the updates every 5 seconds with MQTT.

Typically, when performing processes such as the transition from normal to emergency operation without the rtGovOps framework, golf course managers (or generally operations managers) need to directly interact with vehicle gateways. This usually involves long and tedious tasks such as manually logging into gateways, dealing with device-specific configurations or even an on-site presence. Therefore, realizing even basic governance processes, such as the one we presented above, involves performing many manual and error-prone tasks, usually resulting in a significant increase in

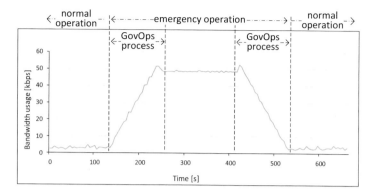

Fig. 5.8: Example execution of operational governance process in the FMS

operations costs. Additionally, in order to be able to have a timely realization of governance processes and consistent implementation of governance strategies across the system, very large operations and support teams are required. This is mainly due to the large scale of the FMS system, but also due to geographical distribution of the governed IoT resources, i.e., vehicles.

Besides increased efficiency, the main advantage that rtGovOps offers to operations managers is reflected in the flexibility of performing operational governance processes at runtime. For example, in Figure 5.8 the execution of the operational governance process took around 2 minutes. In our framework this is, however, purely a matter of operational governance process configuration (naturally with upper limits as we show in the next section). This means that the operational governance process can be easily customized to execute the protocol transition "eagerly", to force the change as soon as possible, even within seconds, or "lazy", to roll out the change step-wise, e.g., 10 vehicles at a time. The most important consequence is the opportunity to effectively manage tradeoffs. For example, executing the process eagerly incurs higher costs, due to additional networking and computation consumption, but it is needed in most emergency situations. Conversely, executing the process in a lazy manner can be desirable for non-emergency situations, since operations managers can prevent possible errors from affecting the whole system.

Figure 5.8 also shows that rtGovOps introduces a slight communication overhead. This is observed in the two peaks at the end of the first process execution, when the framework performs the final checks that the process completed successfully and also when the second process gets triggered, i.e., when the capabilities get invoked on the vehicles. However, in our experiments this overhead was small enough not to be statistically significant. An additional performance-related concern of using rtGovOps is that network latency can slow down the execution of the operational governance process. However, since rtGovOps follows the microservices architecture style, it is possible to deploy relevant services (*API-* and *DeploymentManager*) on Cloudlets [153] near the vehicles, e.g., on golf courses, where they can utilize local wireless networks.

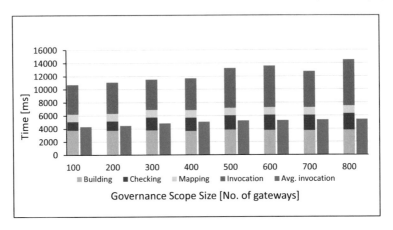

Fig. 5.9: Initial and average time for capability invocation

5.6.4 Results of the Experiments

To demonstrate the feasibility of using rtGovOps to facilitate operational governance
processes in large-scale software-defined IoT Cloud systems, we evaluate its performance in governing approximately 800 vehicle gateways that are simulated in
the previously described test-bed. In our experiments, we mainly focus on showing
the scalability of the two main mechanisms of the rtGovOps framework: (i) capability invocation and (ii) automated capability provisioning. We also consider the
performance of capability checking and governance scope resolution. The reason
why we put an emphasis on the scalability of our framework is that it is one of the
key factors to enable consistent implementation of governance objectives across
large-scale systems. For example, if the execution of an operational governance
process were to scale exponentially with the size of the resource pool, theoretically
it would take infinitely long to have a consistent enforcement of the governance
objectives in the whole system, with a sufficiently large resource pool. The results of
the experiments are the averaged results of 30 repetitions and we have experimented
with five different capabilities that have different properties related to their size and
computational overhead.

Figure 5.9 shows the execution time of the first invocation of a capability (stacked
bar) and the average invocation time of capability execution (plain bar). We notice that
the first invocation took between approximately 10 and 15 seconds and the average
invocation varied between 4 and 6 seconds depending on the scope size (measured
in the number of gateways). The main reason for such a noticeable difference is
the invocation caching performed by rtGovOps. This means that most of the steps,
e.g., building capability image and building the mapping model are only performed
when a capability is invoked for the first time, since in the subsequent invocations
the capability is already in the gateways and the mapping can be done in cache. In
Figure 5.10, we present the average execution time of a capability (as it is observed

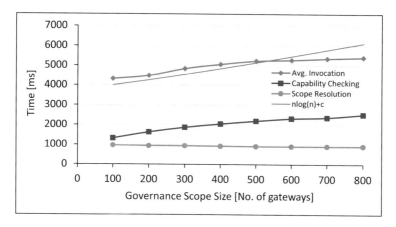

Fig. 5.10: Average invocation time of capabilities against governance scope size

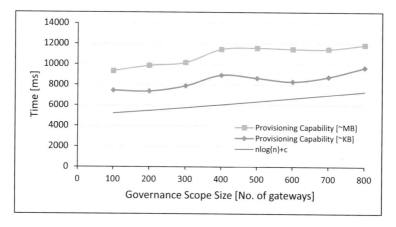

Fig. 5.11: Average capability provisioning duration (push-based strategy)

by an invoking operational governance process on the local machine), average execution of capability checking mechanism and governance scope resolution. As a reference, the diagram also shows a plot of a $nlog(n) + c$ function. We can see that the mechanisms scale within $O(nlog(n))$ for relatively large governance scopes (up to 800 gateways), which can be considered a satisfactory result. We also notice that computational overheads of the capabilities have no statistically significant impact on the results, since they are distributed among the underlying gateways. Finally, it is interesting to notice that the scope resolution time actually decreases with increasing scope size. The reason for this is that in the current implementation of rtGovOps, scope resolution always starts with the global governance scope and applies filters (lambda expressions) to it. After some time Java JIT "kicks in" and optimizes filter execution, thus reducing the overall scope resolution time.

In Figure 5.11, we show the general execution times of the rtGovOps capability-provisioning mechanism (push-based deployment strategy) for two different capabilities. The first one has a size order of magnitude in MBs and the second capability size is measured in KBs. There are several important things to notice here. First, the capability provisioning also scales similarly to $O(nlog(n))$. Second, after the governance scope size reaches 400 gateways there is a drop in the capability provisioning time. The reason for this is that the rtGovOps load balancer spins up additional instances of the *DeploymentManager* and *ImageBuilder*, naturally reducing provisioning time for subsequent requests. Finally, the provisioning mechanism behaves in a similar fashion for both capabilities. The reason for this is that all gateways are in the same network, which can be seen as an equivalent to vehicles deployed on one golf course.

5.6.5 *Discussion and Lessons Learned*

The observations and results of our experiments show that rtGovOps offers advantages in terms of realizing operational governance processes with greater flexibility, and also makes such processes easily repeatable, traceable and auditable, which is crucial for successful implementation of governance strategies. Generally, by adopting the notion of governance capabilities and by utilizing resource agents, rtGovOps allows operational governance processes to be specified with *finer granularity (R3)*, but also gives a *degree of autonomy (R5)* to the managed IoT Cloud resources. Therefore, by selecting suitable governance capabilities, operations managers can precisely define desired states and runtime behavior of software-defined IoT Cloud systems. Further, since the capabilities are executed locally in IoT Cloud resources (e.g., in the gateways), our framework enables better utilization of the "edge of infrastructure" and allows for local error handling, thus increasing system availability and scalability. Further, the main advantage of approaching provisioning and management of governance capabilities in the described manner is that operations managers do not have to worry about geographically distributed IoT Cloud infrastructure nor deal with individual devices, e.g., key management or logging in. They only need to *declare (R4)* which capabilities are required in the operational governance process and specify a governance scope. The rtGovOps framework takes care of the rest, effectively giving a *logically centralized view (R1)* of the management of all governance capabilities. Further, by *automating (R2)* the capability provisioning, rtGovOps enables installing, configuring, deploying, and invoking the governance capabilities in a scalable and easily repeatable manner, thus reducing errors, time and ultimately costs of operational governance.

It should also be noted that there is a number of technical limitations of and possible optimizations that can be introduced in the current prototype of the rtGovOps framework. As we have already mentioned, rtGovOps currently offers limited support for specifying provisioning directives. Additionally, while experimenting with different types of capabilities, we noticed that in many cases better support for dealing with streaming capabilities would be useful. Regarding possible optimizations, in the future we plan to introduce support for automatic composition of capabilities on the device level, e.g., similar to Unix piping. This should reduce the communication overhead of rtGovOps and improve resource utilization in general. In spite of the current limitations, the initial results are promising, in the sense that rtGovOps *increases flexibility* and enables *scalable execution of operational governance processes* in software-defined IoT Cloud systems.

5.7 Summary

In this chapter, we introduced the GovOps approach to runtime governance of IoT Cloud systems in Smart Cities. We presented the GovOps reference model, which defines suitable concepts; and a flexible process to design Smart City and IoT Cloud governance strategies. We introduced the GovOps manager, a dedicated stakeholder responsible for determining suitable tradeoffs between satisfying governance objectives and IoT Cloud system capabilities, and ensure early integration of these objectives with operations processes by continuously refining how the high-level objectives are articulated through operations processes.

Moreover, this chapter introduced the rtGovOps framework that serves as a GovOps reference implementation, providing support for designing and executing operational governance processes. We presented rtGovOps' main runtime mechanisms and enabling techniques that support operations managers in handling two main tasks: (i) perform dynamic, on-demand provisioning of governance capabilities and (ii) remotely invoke such capabilities in the IoT Cloud , via dynamic APIs. We demonstrated, on a real-world case study, the feasibility of the GovOps methodology and framework to facilitate execution of operational governance processes in large-scale IoT Cloud systems.

The initial results are promising in several ways. We showed that the rtGovOps framework enables operational governance processes to be executed in a scalable manner across relatively large IoT Cloud resource pools. Additionally, we discussed how rtGovOps enables flexible execution of operational governance processes by automating the execution of such processes to a large extent, offering finer-grained control over IoT Cloud resources and providing logically centralized interaction with IoT Cloud resource pools. Finally, we discussed how GovOps allows IoT Cloud governance processes to be realized in practice without worrying about the complexity and scale of the underlying IoT Cloud and the diversity of various legal and compliance issues.

Part III
Managing Smart City Social Infrastructure

Preface

The idea of including humans in computational processes has been investigated for decades. Recently, with the onset of portable devices, smart environments and the service-oriented economy, it has once again gained in popularity, promising for the first time to overcome the adoption hurdle. However, most existing approaches model humans as computing elements fitting into existing computational models, as opposed to modeling (computational) processes to fit human elements. In this respect, we are still awaiting a Kuhnian paradigm shift in this area. We are advocates of that shift, and argue that an effective inclusion of humans in socio-technical systems can only succeed if a combination of hard and soft controllability approaches is used. In practice this means redesigning the way we manage the human element in executable processes by relaxing strict constraints to fit the inherent unreliability of humans and embracing the uncertainty that comes with it. The creativity, versatility and sociability of humans should be leveraged to perform runtime adaptations, fix incorrect results and even produce unexpected ones. This is achieved through soft controllability approaches such as incentives, social influence and self-achievement. The concept of the Smart City platform introduced in Section 1.2 represents a prime example of a complex socio-technical system that would benefit from such an approach.

In this part we introduce the core technologies and technological enablers (see Fig. 1.1) for managing the social component of the Smart City platform. We look into the various existing solutions for programmatic management of human participation in socio-technical systems, covering both direct and indirect controllability aspects, and present our own solutions in both domains. These technologies provide technological support for H2H and M2H interactions among stakeholders and are technological enablers for the development of technologies for citizen inclusion and empowerment. In Chapter 7 we introduce the concepts of collectives and collective-based tasks, as an alternative to individually-oriented, role-based task assignment, allowing for self-organization and creation of social fabric between citizens – participants of the collectives. In Chapter 8 we introduce incentives, as a powerful mechanism of soft controllability at scale, and a solution stack for automated incentive management in socio-technical systems.

Chapter 6
State of the Art & Related Work

6.1 Overview of Existing Social-Computing Platforms

Here we present an overview of relevant classes of socio-technical systems and their typical representatives, and compare their principal features with the platform presented in Chapter 7. Based on the way the workflow is abstracted and encoded the existing approaches can be categorized [174] into three groups: *a*) programming-level approaches; *b*) parallel-computing approaches; and *c*) process-modeling approaches.

Programming-level approaches focus on developing a set of libraries and language constructs allowing general-purpose application developers to instantiate and manage tasks to be performed on socio-technical platforms. Unlike our approach, the existing systems do not include the design of the crowd management platform itself, and therefore have to rely on external (commercial) platforms. The functionality of such systems is effectively limited by the design of the underlying platform. Typical examples of such systems are TurKit [102], CrowdDB [57] and AutoMan [15].

TurKit is a Java library layered on top of Amazon's Mechanical Turk offering an execution model ("crash-and-rerun") which re-offers the same microtasks to the crowd until they are performed satisfactorily. The input to TurKit are custom JavaScript files executable on mTurk written by external developers. TurKit manages these files according to the described execution model on behalf of the developer, relieving him of this time-consuming task. While the deployment of tasks onto the Mechanical Turk platform is automated, the synchronization, task splitting and aggregation are left entirely to the programmer. In our solution, the inter-worker synchronization is out of the programmer's reach. The only constraint that a programmer can specify is to explicitly prohibit certain workers from participating in the computations. No other high-level language constructs are provided.

CrowdDB similarly outsources parts of SQL queries as mTurk microtasks. Concretely, the authors extend traditional SQL with a set of "crowd operators," allowing subjective ordering or comparisons of datasets by crowdsourcing these tasks through conventional microtask platforms. As this implies "pay-as-you-go" billing, another construct allows the total price of a query to be limited. From the programming

© Springer International Publishing AG 2017
S. Dustdar et al., *Smart Cities*,
DOI 10.1007/978-3-319-60030-7_6

model's perspective, this approach is limited to a predefined set of functionalities which are performed in a highly parallelizable and well-known manner.

AutoMan integrates the functionality of crowdsourced multiple-choice question answering into the Scala programming language. The authors focus on automated management of answering quality. The answering follows a hard-coded workflow. Synchronization and aggregation are centrally handled by the AutoMan library. The solution is of limited scope, targeting only the designated labor type. None of the three described systems allows explicit collective formation, or hybrid collective composition.

Parallel-computing approaches rely on the divide-and-conquer strategy that divides complex tasks into a set of subtasks solvable either by machines or humans. Typical examples include Turkomatic [93] and Jabberwocky. For example, Jabberwocky's [6] *ManReduce* collaboration model requires users to break down the task into appropriate map and reduce steps which can then be performed by a machine or by a set of human workers. Hybridity is supported at the overall workflow level, but individual activities are still performed by homogeneous teams. In addition, the efficacy of these systems is restricted to a suitable (e.g., MapReduce-like) class of parallelizable problems. Also, in practice they rely on existing crowdsourcing platforms and do not manage the workforce independently, thereby inheriting all underlying platform limitations.

The process-modeling approaches focus on integrating human-provided services into workflow systems, allowing modeling and enactment of workflows comprising both machine- and human-based activities. They are usually designed as extensions to existing workflow systems, and therefore can perform some form of peer management. The three currently most advanced systems are CrowdLang [112], CrowdSearcher[21] and CrowdComputer [174].

CrowdLang brings in a number of novelties in comparison with the previously described systems, primarily with respect to the collaboration synthesis and synchronization. It enables users to (visually) specify a hybrid machine-human workflow, by combining a number of generic (simple) collaborative patterns (e.g., iterative, contest, collection, divide-and-conquer), and to generate a number of similar workflows by differently recombining the constituent patterns, in order to generate a more efficient workflow at runtime. The use of human workflows also enables indirect encoding of inter-task dependencies. The user can influence which workers will be chosen to perform a task by specifying a predicate for each subtask that needs to be fulfilled. The predicates are also used for specifying a limited number of constraints based on social relationships, e.g., to consider only Facebook friends. Even if CrowdLang allows a certain level of runtime workflow adaptability, it is limited to patterns that need to be foreseen at design-time. Our platform differs from both of these systems mostly by extending the support for collaborations from processes known at design-time to fully human-driven, ad hoc runtime workflows.

CrowdSearcher presents a novel task model, composed of a number of elementary crowdsourceable operations (e.g., label, like, sort, classify, group), associated with individual human workers. Such tasks are composable into arbitrary workflows through application of a set of common collaborative patterns which are provided. The workflow is constructed and executed by a custom-designed Node.js framework that first transforms the declarative, task-centric, pattern-based specification into a series of rules and output events, which drive the workflow execution according to the 'event-condition-action' (ECA) paradigm. The actual execution is performed on a third-party crowdsourcing platform (e.g., AMT). The focus of CrowdSearcher's programming model lies in task specification, splitting and subsequent aggregation. This allows a very expressive model but on a very narrow set of crowdsourcing-specific scenarios. This is in full contrast with the more general task-agnostic approach taken by our programming model presented in Chapter 7. The provisioning is limited to the simple mapping "one microtask $\leftrightarrow$ one peer". No notion of collective or team is explicitly supported, nor is human-driven orchestration/negotiation. Furthermore, the system supports only human-based task performers, leaving out the notion of hybridity.

CrowdComputer is a platform allowing the users to submit general tasks to be performed by a hybrid crowd of both web services and human peers. The tasks are executed following a workflow encoded in a BPMN-like notation called BPMN4Crowd, and enacted by the platform. CrowdComputer can be seen as the platform resembling most closely the functionality offered by our platform. However, while CrowdComputer assumes splitting of tasks and assignment of single tasks to individual workers through different "tactics" (e.g., marketplace, auction, mailing list) our platform natively supports actively assembling hybrid collectives to match a task. In addition, by providing a programming abstraction, it offers a more versatile way of encoding workflows.

6.2 Theories of Motivation and Incentives

Intrinsic and Extrinsic Motivation

The fundamental concept related to incentives and rewards is the concept of *motivation*. Motivation has been a topic of interest in psychology for decades, with different theories emerging in different epochs. The resulting corpus of research led to the commonly accepted view of today [150], where motivation is classified into two major types – *intrinsic* and *extrinsic*.

Intrinsic motivation is described as the driving force attracting individuals to perform an activity for the inherent satisfaction associated purely with the act of performing that activity. Extrinsic motivation, on the other hand, is the driving force pushing individuals to perform an activity in which they find no inherent interest or satisfaction, but which is associated with external rewards substituting the missing inherent satisfaction.

Intrinsic motivation is powerful and stable, often associated with curiosity, creativity, competitiveness, playfulness and volunteering. However, it can also be highly dependent on different social and environmental factors. On the other hand, extrinsic motivation is exerted directly by an intervention strategy (e.g., incentive), making it more controllable, but also more volatile, as its effects are gone when the intervention is absent. Extrinsic motivation is important for performing activities which individuals consider important and necessary but not inherently enjoyable (e.g., learning, working). The different motivation types, and how they influence human behavior, are introduced and researched as part of Self-Determination Theory (SDT) [39].

From the operational aspect (which is of particular interest to us), the two motivation types can be treated through the notion of *reward*. While the concept of reward is inherent to the definition of extrinsic motivation, in case of intrinsic motivation the reward can be considered as providing the opportunity to perform an activity or get better at it. The same rationale is also adopted by *Operant Theory* [1]. This operationalization allows us further to define *incentivization* as the process through which motivation is fostered by application or provisioning of rewards.

As a practical example, this means that by incentivization we consider both a promise of payment of a monetary reward (extrinsic motivation), as well as providing an opportunity for an amateur astronomer to voluntarily participate in a citizen-science galaxy identification program (intrinsic motivation). In both cases incentivization influences the individual's *locus of control* – in the former case the locus is external; in the latter the locus is internal.

Principal-Agent Theory

Historically, there has been much more commercial interest in investigation of extrinsic intervention strategies. The practical concept of *incentives* appeared together with the division of labor. Delegating productive tasks to others (workers) meant it was necessary to make sure that workers, pursuing own interests, did not work against the owner's interests. Incentives served primarily this purpose – to align the interests of the owner and the workers. This meant that the most important extrinsic operational models and theories were originally developed by economists. An overview of the historical development of the notion of incentives and rewarding in economic thought can be found in [95].

The predominant model of incentives used in economics was set out in the *Principal-Agent Theory* (also known as *Agency Theory*) [16, 95]. The theory introduces the role of *principal*, corresponding to a manager in a traditional firm, who delegates tasks to a number of *agents*, corresponding to employees under his supervision. It is assumed that the agents seek to minimize their effort and risks while maximizing their compensation. The principal wants to minimize the costs of the agents' labor and maximize profits. Therefore, their interests diverge.

Every agent is unique, possessing unique qualities and properties. For the same task different agents will put in different levels of effort, and will value that effort at a different cost[1]. Additionally, every agent can, unobserved by the principal, perform certain actions that go against the principal's interests. The theory implies that the

---

[1] All the costs and prices, as well as amounts of incentives are expressed as numerical quantities. It is assumed that these numerical values include and represent also any other properties that the

agents know their personal values for these properties (*adverse selection*) and know their intentions to commit hidden actions (*moral hazard*). The fact that they remain secret to the principal is called *information asymmetry* or the *information gap*.

We assume that the principal knows just the statistical distribution of these values. If the principal knew entirely the private information (*signals*) for every agent, then each agent could be offered the ideal *contract* from the principal's point of view, i.e., paying him just as little as it takes for him to perform the work with the given effort. However, this not being the case, some agents get overpaid while others get underpaid, effectively inducing them not to accept working on the task. The principal wants to know and measure as many agent signals as possible, because the more insight he gets into agent's capabilities and behavior, the more chances he has of setting up a better contract and maximizing his profit (*Informativeness Principle*). So, the principal offers the agent an incentive to disclose part of this information in order to compile a better-suited contract.

Incentives are an additional expense for the principal. However, if, based on the new information, the principal can offer better contracts and consequently make more profit by filtering out and motivating quality agents, then the investment in the incentives will pay off. Therefore, the incentive designer is faced with an optimization problem that involves human agents, whose individual behaviors cannot be foreseen. A way to solve this problem lies in assembling an incentive strategy comprising a number of simple incentives whose effects on the majority of agents can be predicted closely enough, and then adapting the strategy based on concrete, context-specific feedback. In a traditional company, this would mean that a manager would offer a combination of wage increases, free days, promotions, bonuses and other benefits to the workers that achieve higher levels of some wanted property (e.g., productivity, quality, knowledge, leadership). Increased expenses for the principal are compensated not only by increased productivity, but in fact much more by the selective effects of the incentives [98]; by investing in incentives the management gains the knowledge of which workers can produce more value to the company and therefore should be stimulated to stay in the company.

The fundamental difficulty when applying the theory in practice lies in precisely defining and subsequently measuring the different qualities of agents and their performance (signals). As we previously mentioned, the Informativeness Principle states that each contract should be designed with as many signal measurements taken into consideration as possible. As it is in the interest of the agents to keep some signals private to them, measuring those signals becomes the major obstacle. In practice, working involves performing a lot of complex and interrelated tasks, and often collaborating and depending upon different people, so the principal problem translates into the inability to effectively assess the quality of a particular worker's performance in a dynamic and complex environment due to the impossibility of quantifying and measuring all of the signals. This is even more accentuated in social-computing environments, where contracts (in the sense of Agency Theory) are more persistent than the signal sets that need to be considered, calling for frequent contract

---

agents and principal value, such as risk, free time, gratification due to pleasant working environment, personal satisfaction, prospect of promotion, fear of dismissal, etc.

and incentive adjustments. Additionally, this also causes a number of behavioral responses (*dysfunctional behavior*) in agents meant exclusively to increase rewards while damaging overall performance levels.

The agency theory implies a fully rational, self-interested agent, who always acts in his/her best interest (so called *homo oeconomicus*). The incentive is always monetary and acting on extrinsic motivation. In practice, this is not always a sufficiently accurate model. For this reason, additional *decision-making theories* and multidisciplinary frameworks were developed, taking into consideration various determinants of behavior, including additional factors of intrinsic motivation, environmental and social factors, and assuming agents with bounded rationality. In [51, 178] a comprehensive overview of different incentive theories and decision-making frameworks is presented. However, while providing more realistic and nuanced behavioral models, these frameworks are less suited for technical abstraction and exploitation. The agency theory remains, therefore, an important theory that is practically applicable in appropriate working environments.

Both the agency theory, as well as the more complex decision-making frameworks state that the effect of incentives on an agent's behavior is exhibited dually, through *selective (sorting) effects* and *performance effects*. Selective effects are defined as the act of revealing more precise details of an agent's qualities, shortcomings and performance parameters through monitoring the application of incentives or through agent self-selection. Performance effects are changes in an agent's performance caused by behavior modified through application of incentives.

Depending on the incentive and the application context one effect type may be more expressed than the other. Also, one type may be valued more than the other by the principal. For example, in piecework productive activities performance effects are usually valued more, while in engineering disciplines discovery of creative workers may in the long run be more profitable than the rewarding of the currently more productive ones.

Efficacy of Incentives

Although it sounds a commonplace that offering monetary rewards to someone should gratify that person and induce him/her to perform better, different research efforts demonstrated through empirical studies that it may not always be the case. For example, in [58] the authors empirically conclude that in some cases the monetary rewards actually decrease intrinsic motivation. On the other hand, in [97] for example, we encounter strong empirical evidence that in specific cases monetary rewards do significantly increase performance.

However, all the studies conclude that, depending on the environment, there always exist types of incentives that can provide the necessary motivation. With some simple, repetitive tasks, paying for performance increases productivity [97, 109, 106]. Professionals that value the humanistic impact of their work (volunteers, community workers, firemen, doctors, scientists, etc.) may be intrinsically motivated by having the positive contribution of their work to society shown [70, 74]. In companies with lengthy and complex tasks promotions and/or team-based rewards are effective [177, 85]. Finally, as mentioned before, sometimes the sorting effects of incentives are much more useful to the principal than possibly moderate performance effects.

The expertise on the expected effects of particular incentives is based on empirical data usually formulated as very general claims about behavior of agents under certain conditions and then proven by different empirical methods (see [51, Ch. 6] for an overview). However, as in many areas dealing with human behavior, absolute and quantifiable incentives cannot be given in advance, but rather must be adapted for a given collaborative environment after a careful study of the context and the habitual/cultural background of the participants.

While choosing appropriate incentive strategies and adapting them to fit specific types of labor may prove a challenging task, it is a conclusive fact that incentives can exhibit considerable selective or performance effects on workers, corroborated in practice by the fact that most traditional businesses employ incentive schemes [51, Ch. 1].

6.3 Incentive Management in Computer Science

Most related work in the area of rewarding and incentives originates from economics, organizational science, psychology, and applied research, mostly for military purposes. It can be used to classify and substantiate the basic rewarding approaches, and expected outcomes, and to simulate the responses to our incentive strategies. There is only a small number of computer science papers that treat the topics of incentives and rewarding, usually within particular application contexts (e.g., peer-to-peer networks, agent-based systems). However, to the best of our knowledge, no other computer science work treats the topic in a comprehensive manner. In fact, most papers completely disregard the existing theoretical foundations of incentives, and are concerned with solving only the particular problem, as we will show in the rest of this section. The work [178] is a notable exception, discussing incentives designed to motivate participation in different social computing platforms and relating them to the leading behavioral theories, and presenting a vision for future developments in this research area.

In [152] the aim is to maximize P2P content sharing. Therefore, they define roles of (content) forwarders and receivers. The forwarder gets a reward when the receiver reacts in some way to the content being forwarded. They then define the prices of forwarding and receiving actions, and assign incentives based on that. Many other papers similarly identify certain behavioral patterns and develop particular solutions to prevent unwanted behavior or enhance existing algorithms to optimize certain metrics ([83]).

The paper [55] discusses ways of modeling and implementing adaptable agent-based systems. Each agent can be modified by adding or removing modules that make up the agent. The cause of agent adaptability is usually a role change within an organization.

In [102] the authors try to determine the quality of work achieved when a task is done iteratively compared to when it is done in parallel. The difference is that in iterative processes (when applicable), workers are shown previous attempts by other workers, which can influence their work in positive or negative ways. They conduct experiments with real workers on Amazon Mechanical Turk, and prove statistically that, when applicable, the iterative approach yields better (more accurate) results. The quality of the work in their experiments is quantifiable, or voted by the crowd. This is an important finding, since it justifies the choice of iterative execution model that we adopt.

In [109] two basic findings seem robust, and can be used as general conclusions when modeling behavior of contributors: "First, that paying subjects elicited higher output than not paying them (where increasing their pay rate also yielded higher output); and second, that in contrast to the quantity of work done, paying subjects did not affect their accuracy. Although surprising, this latter result may be related to an "anchoring effect" in that subjects' perception of the value of their work was strongly correlated with their actual pay rate."

In [106] the authors compare the performance between paid and volunteering workers by running experiments on well-known commercial platforms (Amazon Mechanical Turk, Zooniverse and Planet Hunters) and analyze different reasons for improved or worsened performance. Interestingly, although the experiments bear much resemblance to psychological experiments intended to measure intrinsic motivation in the context of SDT, no reference to those experiments is made.

In [180] the authors investigate whether self-governing and self-coordinated human teams (without a centralized authority) can be stable if individual members of such teams follow appropriate rules. In [188] the authors seek to maximize the extent of a social network by motivating people to invite others to visit more content (i.e., give a contribution measured in number of pages), and evaluate a number of concrete rewarding schemes (e.g., Dynamic Differentiated Rewarding Scheme). In [72] the authors analyze two commonly used approaches to detect cheating and properly validate crowdsourced tasks. In [15] the focus is on pricing policies that should elicit timely and correct answers from crowd workers. The paper [78] examines which psychological and monetary incentives are used to lure social network users to click on malicious links. In [140] the authors analyze how incentive schemes relying on peer voting can influence the decisions of workers on a crowdsourcing platform.

The major limitation of these research approaches (see [4]) is that the findings are applicable only for a limited range of activities, considered as conventional crowdsourcing tasks, such as image tagging, multiple-choice question answering, text translation or design contests. Furthermore, differences in cultural background [66] can also skew the findings. However, the results of the listed papers, taken together, can provide some generalizable findings that need to be taken into account when designing an incentive management system. For example, the finding that the transparency of actors and processes in a socio-technical system will likely improve the overall performance [77] for us translates to the requirement of portability and transparency of incentives. The findings of [64] indicate that for performing more intellectually challenging tasks smaller groups of expert workers may be more effective than web-scale crowd collectives. Again, this is in line with our motivation of supporting novel socio-technical systems employing smaller teams of experts rather than large anonymous crowds only. Similarly, the aforementioned difference of effectiveness in different cultural backgrounds maps to the requirements of usability and expressiveness to offer to incentive designers a tool for quick adaptation of general incentive mechanisms into locally effective versions.

Chapter 7
Programmatic Management of Human Coordination and Collaboration Activities

Complex coordinated activities involving citizens, their devices and various services (Section 1.4) represent one of the key defining properties of the described Smart City vision and an enabler of novel societal and business values. These activities can be orchestrated and coordinated: centrally (by platform), distributively (by citizens), and hybridly (by platform and citizens). How effectively such activities can be managed depends on both the complexity of the performed tasks and the number and kind of participating actors.

Centrally coordinated activities are suitable both for large-scale collaborations if the performed tasks are simple (e.g., crowdsourcing microtask platforms) and for highly complex tasks, albeit with well-defined execution steps and a relatively low number of participating actors with precisely determined roles (e.g., workflow systems). The latter approach is particularly suitable in conventional business environments, but less so in a general Smart City environment where processes involving citizens cannot always be precisely defined.

Distributively coordinated systems supporting execution of complex joint activities are extremely difficult to design and develop. The current state of the art is nature-inspired swarm/organ systems, where individual self-sufficient units form around collective goals[1]. However, in the context of the envisioned Smart City goals, such systems are unable to support the necessary complexity of processes.

Hybridly orchestrated systems combine the advantages offered by the central point of control (the platform) to impose the overall choreography and manage trust, scale and execution constraints, while delegating the complexity of determining the low-level execution steps and actual actors to the participants (citizens). Since the workflow can be determined at runtime by the participants this drastically reduces the complexity of the platform, and since the human participants are expected to self-organize and agree on execution steps, such an approach allows complex and creative ad hoc problems to be solved. However, this comes at a price: finding the participants to perform a task, communicating the task goals to the participants, and having participants reach an agreement on the execution steps are all phases

---

[1] See http://www.focas.eu/

© Springer International Publishing AG 2017
S. Dustdar et al., *Smart Cities*,
DOI 10.1007/978-3-319-60030-7_7

of the task execution with a high risk of failure. Nonetheless, these phases mimic a human-centric approach to solving problems, where a team of people is formed to solve a problem and given a free hand to find the solution under a best-effort assumption. We believe that this approach is suitable for a Smart City environment, and describe in this chapter a prototype of such a system – the *SmartSociety platform*.

7.1 Research Context

The Smart City platform described in Section 1.2 presents a unified view of the necessary core functionalities for the management of the overall Smart City infrastructure, including the social component. In this chapter, we present the *SmartSociety* platform, which was the result of four years of research[2] in the area of Collective Adaptive Systems (CAS) within the EU FP7 SmartSociety[3] project. The platform fits well into the overall Smart City vision presented here, as it allows the programmatic management of collectives – ad hoc self-assembled teams of humans supported by software services, performing tasks of arbitrary complexity both in digital and in physical domains (such as a city). As such, it can be considered an integral part of the overall Smart City platform.

7.1.1 The SmartSociety Platform

Efficient management of Smart City infrastructure implies blurring the line between human and machine infrastructural elements whenever possible and considering them under the generic term *peer*. A peer is an entity providing different functionalities under different contexts, provisioned under a service model [193], participating in *collectives* – persistent or short-lived teams of peers, representing the principal entity performing an arbitrary task. Peers and collectives embody the two fundamental properties of the CAS vision: *hybridity* and *collectiveness*, offered as inherent features of CAS systems, such as the SmartSociety platform.

The SmartSociety platform is an open source software toolkit[4] intended for use by:

1. *Users* – external human/software clients who need a complex collaborative human-machine task performed;

---

[2] DISCLAIMER: Although the majority of the material presented in this chapter relies on the authors' own work within the SmartSociety project, in order to present a complete picture, the overall description of the platform design contains descriptions of the components developed by SmartSociety partner institutions.

[3] www.smart-society-project.eu

[4] https://gitlab.com/smartsociety/

2. *Peers* – human or machine entities providing Human/Web Services (possibly for compensation).

The platform acts as intermediary between the two user types, trying to align their interests and provide them the following functionalities:

- For users: *a)* a task execution environment; and *b)* workforce management functionality.
- For peers: *c)* a collaboration environment; and *d)* fair working conditions.

The intended platform usage context foresees human peers registering their profiles with the platform and enlisting to perform different professional activities and provide services. These activities can take place both in the digital as well as in the physical world (e.g., sharing a car ride after agreeing through a digital service). The platform uses the profile data to locate and engage peers in different collaborative efforts. Peer engagement is transparent with respect to the working conditions – peers know in advance the conditions under which they are required to provide their services, how the effort will be monitored/assessed, as well as what kind of compensation (or penalty) awaits them. In case of human peers, the platform asks for explicit approval, enabling the peer engagement under a short-term contractual relationship. In case of a software peer, the services are contracted under conventional service-level agreements (SLAs).

Once the platform has located appropriate peers to perform a task (computation), a *collective* is formed. A collective is composed of a collaborative environment and a team of peers assembled for to perform a specific task.

The collaborative environment consists of a set of software communication and coordination tools. For example, the platform is able to set up a predefined virtual communication infrastructure for the collective members, provide access to a shared data repository (e.g., Dropbox folder) [193], and provide tools for coordination of the necessary activities to be performed by the collective's members (e.g., Doodle).

The collective assembly and dissolution may be executed either by the peers or by the platform. In either case, the platform enforces specific negotiation and composition protocols. In this way the platform is able to fully manage the collective lifecycle, and use collectives to perform collective tasks in the context of SmartSociety *platform applications*. During the task execution various incentives and rewards [161] may be applied to stimulate the collective's effort and retain the peers. After finishing the task, collectives may be dissolved, and the reputation and other metrics of the member peers are updated as a testimony for future participations.

The platform applications are the means by which the Smart City stakeholders tap into the social infrastructure resource pool. They are an embodiment of the added-value services that can be built on top of the Smart City platform (Figure 1.2). This is possible because platform applications contain arbitrary business logic permitting their use in various business/societal domains, but rely on the platform's API to leverage the citizens' cognitive and physical capabilities in a uniform way. The platform handles the citizens' privacy and reputation, guarantees fairness in negotiations, agreements and rewarding, and stipulates universal legal conditions.

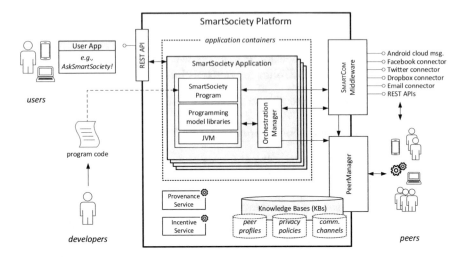

Fig. 7.1: Positioning programming model components within the overall SmartSociety platform architecture

A *user application* communicates with the corresponding platform application. For example, as shown in Fig. 7.1, a mobile user application SmartShare contacts the corresponding SmartSociety platform application. Note that the same person can at the same time play the role both of a user and of a peer of a platform application – for example, in a ride-sharing application, a person requests a ride from the platform (as a user), but then takes part in the ride (e.g., as a driver) and thus plays the role of the peer by providing a service for the platform. Similarly, the same person can participate as a peer in different collectives, in the same or different platform applications concurrently, represented by different peer profiles.

7.1.1.1 Platform Architecture & Functionality

A simplified, high-level view of the SmartSociety platform architecture is presented in Fig. 7.1. The architecture is designed to be fully distributed and scalable. The rectangular boxes represent the key platform components that may be deployed distributively, as all components expose (private or public) RESTful APIs. The principal component-interoperability channels are denoted by double-headed arrows in the figure. Communication with peers is additionally supported via popular commercial protocols to allow broader integration with existing communication software and allow easier inclusion of peers into the platform.

User applications contact the platform through the REST API component. All incoming user requests are served by this module, which verifies their correctness and dispatches them to the appropriate platform application that will be processing and responding to them. The platform applications run sandboxed in appropriate containers, allowing the applications to be deployed on different (virtual) machines. The first time a platform application is run the container will take care of informing the *Peer Manager (PM)* component to set up (register) the appropriate application, peer and collective profiles required by the application. The container will also request from the Peer Manager a set of permissions for accessing and manipulating sensitive private data of peers. The Peer Manager can be described in short as the central data store of the platform, managing all peer and application information and allowing privacy-aware access and sharing of the data among platform components. In practice, the platform application is a Java application making use of the SmartSociety platform's programming libraries, allowing the developer to execute collective-based tasks on the platform. Each platform application features a dedicated *Orchestration Manager (OM)* component. The OM is the component in charge of preparing and orchestrating collaborative activities among peers. Performing these functionalities requires the OM to use the Peer Manager and SMARTCOM Middleware components.

Principal Components

Peer Manager

The *Peer Manager* provides the central data store that maintains and manages information about human- or machine-based peers in a privacy-preserving framework. Concretely, the Peer Manager provides the following functionalities: *a*) A mechanism to manage peers' information (using profiles) that accounts for heterogeneous peers; *b*) A semantic peer search functionality; and *c*) A model for enforcing advanced privacy mechanisms.

To effectively manage peer profiles across different applications, the Peer Manager builds upon the notion of an *entity-centric semantic enhanced model* [62] that defines an extensible set of entity schemas providing the templates for an attribute-based representation of peers' characteristics. The concrete meaning of schemas is specified by mapping single elements (i.e., types of entities, names of attributes and their values) to concepts from an underlying ontology that is also part of the same model, thus allowing reasoning over peer's properties as well as the implementation of semantic-enhanced services. The basic set of schemas/templates can be easily extended to support new application-specific attributes, allowing efficient definition of new peer types. The adaptability is provided by enabling search and information-sharing services to work over the new types in a way that is transparent to the rest of the platform.

By leveraging the semantic search approaches described in [63] the PM's search functionality allows relevant peers and collectives to be located based on a set of attribute constraints even when they are described using different terminology in their profiles. Queries can specify arbitrarily complex semantic operations and constraints on attributes. The semantical search is one of the enabling factors for the overall hybridity property of the platform, because it allows queries originating from human

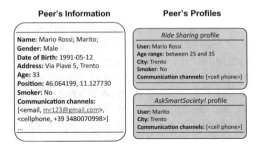

Fig. 7.2: Simplified example of a peer with multiple profiles. Each profile is revealed to a different application

peers to be interpreted into a tractable set of constraints, thereby alleviating the semantical differences that are inherently present when dealing with humans.

Additionally, the Peer Manager defines a *privacy protection model* that pays special attention to different privacy principles enacted by EU Data Protection Directive 95/46/EC[5] affecting storage and processing of personal data. Specifically, the model defines privacy regulations and considerations described in [68], such as purpose specification and binding, that are enforced upon search queries. This means that in different usage contexts the peer profiles will reveal only partial or (semantically) obfuscated information, used for replying to specific information requests, thus enforcing data minimization. Fig. 7.2 shows a simplified example of a human peer subscribed to participate in two platform applications: a ride-sharing application and a Q&A application, revealing different information (by using different profiles) in each case. This allows, e.g., a human peer to reveal his age range (as a way to obfuscate the exact date of birth) when participating in a ride-sharing collective, while the same information is completely hidden when participating in a question-answering collective.

Communication Middleware

SMARTCOM is communication and virtualization middleware used as the primary means of communication between the platform and the peers. Although tightly integrated into the platform, SMARTCOM is designed as an independent component that can be used with similar CAS platforms. Apart from performing functionalities typical of conventional service buses (e.g., message transformation, routing, encryption, authentication) the distinguishing novelty of SMARTCOM is its native support for virtualizing collectives [193]:

- Hiding the complexity of communication with a dynamic collective as a whole and passing instructions from the CAS platform to it, making the collective a first-class, programmable entity (DR3);

---

[5] http://eur-lex.europa.eu/legal-content/EN/ALL/?uri=CELEX:31995L0046

- Making the human vs. machine distinction transparent during the communication, by interpreting/translating the messages for different peer types and delivering them to peers through different communication channels/protocols, in accordance with the peer's preferences (DR2);
- Allowing concurrent participation of peers in different collectives, acting as different service units with different SLAs, delivery and privacy policies.

SMARTCOM's design and functionalities are described separately in Section 7.2 as it is one of the key components contributing to the hybridity of the future Smart City platform.

Orchestration Manager

The *Orchestration Manager (OM)* is responsible for the following functionalities:

- *Composition* – Generating possible *execution plans* to meet user-set constraints and optimize wanted parameters.
- *Negotiation* – Coordinating the negotiation process among human peers leading to the overall agreement and acceptance of the suggested execution plan by the participating peers.
- *Execution* – Monitoring the execution and enforcing the selected execution plan during runtime.

The OM works in an asynchronous loop reacting to events of new (users') task requests and (peers') participation requests. Upon each event the OM computes the set of feasible execution plans associated with one or more requests. Plans are constructed by solving a high-level combinatorial or constraint satisfaction problem (see [41]).

For example, in a ride-sharing scenario, drivers post the rides (task requests) and passengers express participation requests (also a type of task requests). Although passengers may be flexible to take a ride in different time intervals during the day, an execution plan can contain only a time interval fitting every participant in the riding collective associated with that plan. Other constraints may need to be considered, such as the capacity of the vehicle; or trade-offs, such as choosing between the optimal route vs. the route that accommodates more participants. In such a setting, each new/altered request can lead to creation/invalidation of multiple plans – e.g., a number of passengers who submitted a participation request could not have been part of any execution plan until a driver submitted the matching ride offer. When the ride is finally offered, multiple possible plans are generated with different passenger collectives, and only one plan can (in this case) be ultimately realized. Furthermore, if at any time the driver cancels the ride, all plans need to be invalidated. Conversely, a matching ride offer by another driver creates a different set of execution plans, opening up the possibility for passengers to concurrently consider and negotiate about participating in different rides, but ultimately choosing only one. Once new plans are generated, the participants in the tentative collectives associated with each plan can negotiate among them, thereby deciding whether the candidate solution provided by the OM is acceptable and the actual execution can take place. The OM mediates the negotiation process based on the selected *negotiation protocol (pattern)*. The OM uses SMARTCOM to enact the negotiation protocol, i.e., to dispatch appropriate offers, accepts, rejects and agreed plans.

The described OM functionalities are fundamental to enable the human-driven collectiveness, i.e., workflows where the order of activities is not prescribed, but is instead determined at runtime, based on the preferences and capabilities of the human peers interested in performing the task.

Programming and Execution Model

Collective management and task execution are enabled by the platform's *Programming and Execution Model* (for short, Programming model). The Programming Model is the umbrella term for a set of concepts and the associated communication, interaction, coordination and execution models that abstract the fundamental CAS notion of hybrid collectives for the developer, and provide the means, in terms of appropriate high-level language constructs, to manipulate them, and ultimately execute them on top of the SmartSociety platform.

The fundamental novelty of the programming model is achieved by replacing the individual peer with the notion of *hybrid collective* as the central processing unit. This allows the responsibility for associating tasks with peers and orchestrating the collaboration to be shifted from the developer of the application running on the platform to the participating peers themselves. The platform then merely coordinates the participating humans in a human-driven collaborative process. At the same time, the coordination management also shifts – from a design-time, platform-administered one (requiring a predefined workflow) into a runtime, human-driven coordination management with unpredictable workflow. All coordination and execution happens encapsulated within the introduced notion of collective transparently to the developer, allowing him to focus on *what* needs to be performed, and leaving the *how* part to the platform (a declarative approach).

This opens up the possibility of supporting a whole spectrum of collaborative patterns (labor models), ranging from *on-demand* to *open-call*. In open-call collaborations tasks are published on the platform, and peers are motivated to apply and negotiate to take part in execution of the task. The negotiation and execution are completely human driven. In on-demand collaborations, for each input task the platform tries to locate or provision peers/collectives that are capable of performing the task with respect to the given input constraints (e.g., [163]).

The Programming model libraries represent the heart of the platform. Section 7.3 explains the design and operating principles of the Programming model, showing how it puts to use the principal platform components.

7.2 Communication and Virtualization Middleware

The SMARTCOM middleware was designed[6] to perform the bulk of the general collective communication and virtualization tasks for socio-technical platforms. Here we present only the functionality and design traits relevant to the context of the Smart City platform:

1. Message queuing, routing, transformation and delivery of messages between citizens (peers) and the platform.
2. Support for different messaging formats to support coupling with different platforms but also allowing direct communication with human participants through popular protocols (e.g., email, Android notifications, Twitter).
3. Privacy and anonymity isolation layer.

The full description of the middleware's architecture and functionality is provided as a separate technical report[7].

7.2.1 Architecture

Figure 7.3 shows the internal architecture of the SMARTCOM middleware. The primary function of the middleware is exchange of messages between *peers* and the *executing platform* (e.g., Smart City or Smart Society platform), as well as among peers themselves. The term *peer* is used to denote both human entities (i.e., citizens/workers) and software entities (e.g., external Web Services) that act as communication endpoints (senders/receivers of messages). The executing platform is the software entity performing computational processes involving peers, for which SMARTCOM provides communicational support. The term *collective* is used to denote a set of peers requiring multicast routing and delivery at a given time. For example, in an incentive context, a collective can represent a team of workers who need be contacted concurrently via different communication channels and protocols to deliver informational or motivational messages.

The executing platform passes the messages intended for collectives to SMART-COM (i.e., to the *Communication Engine* component) through a public API. The task of the Communication Engine is to virtualize the notions of peers and collectives to the executing platform, determine the recipients and delivery routes and instantiate an 'adapter' to perform the delivery. The term *adapter* denotes the middleware component in charge of handling the communication.

---

[6] DISCLAIMER: Parts of the SMARTCOM design and implementation were co-authored by Dipl.-Ing. Philipp Zeppezauer under the authors' co-supervision. The results were published in the joint publications [193, 192]. Some material from the cited joint publications is used here.

[7] https://github.com/tuwiendsg/SmartCom/blob/master/doc/technical-report.pdf

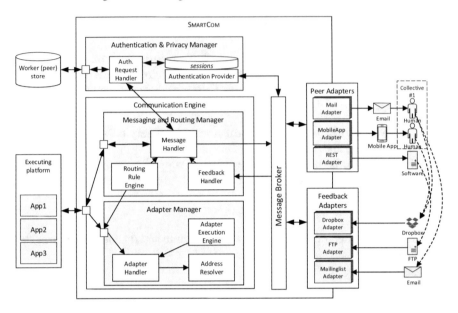

Fig. 7.3: Internal architecture of SMARTCOM middleware

7.2.2 Messaging and Routing

All communication between the peers and the platform is handled asynchronously using normalized messages. A queue-based *Message Broker* is used to decouple the execution of SMARTCOM's components and the communication with peers. SMARTCOM supports unicast as well as multicast messages. Therefore, multiple peers can also be addressed as collectives and the SMARTCOM will take care of sending the message to every single member of the collective.

The *Messaging and Routing Manager* (MRM) is SMARTCOM's principal entry point for communication with the platform. It consists of the following components: 1) The *Message Handler* takes incoming messages from the platform and transforms them into an internal representation, sending it to the receiver via a determined peer output adapter. If the receiver of the message is a collective, it resolves the current member peers and their preferred communication channels, determining a set of output adapters to use; 2) The *Routing Rule Engine* then determines the proper route to the peers, invoking the Adapter Manager to instantiate appropriate adapters in order to complete the route, if needed; 3) The *Feedback Handler* waits for feedback messages received through feedback (input) adapters and passes them to the Message Handler. Afterwards they will be handled like normal messages again, and re-routed where needed, e.g., back to the platform. A *route* may include different communication channels as delivery start/endpoints. Figure 7.4 shows the conceptual overview of SMARTCOM's routing. For each message the route will be determined by the Routing Rule Engine using the pipes-and-filters pattern, determining the route based on the message properties: receiver ID, message type and message subtype, with decreasing priority. Note that there may be multiple routes per message (e.g., a single peer can be contacted using a mobile app and email concurrently).

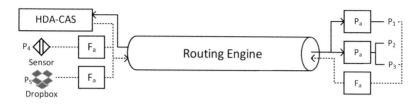

Fig. 7.4: Messages are routed to Output Adapters (P_a) which forward the messages to the corresponding Peers (P_1 to P_5). Feedback is sent back by human peers, software peers (e.g., Dropbox) and sensors using Input Adapters (F_a). The platform (HDA-CAS) can also send and receive messages

Message Structure and Types

The message structure is similar to the FIPA ACL Message Structure [52], but some properties have been removed and others added to fit the requirements of SMARTCOM.

Each message consists of several mandatory and optional fields. The most important fields of a message are the Id of the message, the sender, the type and subtype. These and further fields are discussed and described in Table 1. Listing 7.1 outlines a simple message containing instructions for a task in the JSON format.

```
1  {
2    "id": "2837",
3    "type": "TASK",
4    "subtype": "REQUEST",
5    "sender": "peer291",
6    "receiver": "peer2734",
7    "conversation-id": "18475",
8    "content": "Check the status of system 32"
9  }
```

Listing 7.1: Example message with instructions for a task

Field	Description
Type	This field defines the high-level purpose of the message (e.g., control message, input message, metrics message). This field is especially important for the routing of messages within the system
Subtype	This field is defined by the component that is in charge of the message (i.e., it is component-specific). The subtype combined with the type of the message defines the purpose of the message. The subtype can also be used by programmers of applications to define custom message types for their application
Message-ID	A global unique identifier is assigned to every message within the system by the Messaging and Routing Manager
Sender-ID	The sender-ID specifies the sender of the message (can be a component, peer, etc.). Sender-IDs are unique within the system. Sender-IDs are either predefined in case of an internal component or are assigned by the platform component
Receiver-ID (o)	The receiver-ID specifies the receiver of the message (can be a component, peer, collective). Can also be empty if the receiver is not clear
Conversation-ID (o)	Denotes the system identifier for the conversation. This identifier can be used by platform components to map the message to the actual execution instance of an application. For example: application A is executed twice at the same time: A_1 and A_2. The conversation-ID is used to associate the messages with the right execution A_1 or A_2. If there is no conversation (e.g., for internal messages), the conversation-ID can also be empty
Content (o)	Defines the content of the message including instructions and data that are needed to execute the message. This can be empty in case of simple messages (e.g., acknowledge messages)
TTL (o)	Time to live. Defines a time interval in which a message is valid. For example: a peer has one hour to post pictures in a folder of an FTP server, after this time SMARTCOM stops looking for pictures in the folder and creates an error message if there are no pictures
Language (o)	Denotes the language of the message. This can be a natural language, such as English or German, or a binary format. The initial intention of this field is logging and debugging purposes. In future versions a translation service could be introduced that makes use of this field
Security-Token (o)	The security token can be used to guarantee the authenticity of messages or to encrypt the content of messages
Delivery-Policy (o)	Specifies the delivery policy of the message. This field can be used to specify whether the sender wants an acknowledgment in case of a successful delivery of the message
RefersTo (o)	This field can be used to specify that this message refers to another message

Table 1: Structure of messages. Optional fields are marked with (o)

Predefined Messages

These messages are needed for special purposes, such as authentication, or to indicate specific behavior (i.e., an acknowledged message) or exceptional cases and errors. The following sections describe these predefined messages and define their intended usage in the system. The subtypes of the messages are defined in the corresponding rows within brackets and in capital letters.

Control messages are exchanged within SMARTCOM and are exposed to the application. Control messages are always indicated by the message type *CONTROL*. Their intention is to indicate specific control behavior (e.g., acknowledgement of a message) or exceptions during the communication. Table 2 presents the various subtypes.

Message	Description
Acknowledge (*ACK*)	This message is sent by the output adapter if the message has been successfully sent to the peer. Note that this does not imply the peer's acceptance of the contents of the message, but is used to implement functionalities such as read receipts. This message is not sent if the programmer requires a fire-and-forget sending behavior (i.e., she doesn't care if it actually has been delivered)
Error (*ERROR*)	An error message that indicates a generic error. This message is handled based on the routing rules
Communication Error (*COMERROR*)	This error message indicates an error during the communication. This is reported to the sender of the initial message
Timeout (*TIMEOUT*)	This message indicates that a timeout has appeared in the system and that the message couldn't be delivered in time or there was no response within a certain time

Table 2: Predefined subtypes of Control Messages

Authentication messages are used to perform authentication of a peer in the system and provide him with a security token that is valid for a specific time period. Such messages are handled by the Authentication Manager which interacts with platform components to verify the identify of a peer. Authentication messages always have the type *AUTH*. AuthenticationRequest messages are sent by peers to the system whereas the other three messages (AuthenticationResponse, AuthenticationFailed, AuthenticationError) are sent back from SMARTCOM to the peer. Table 3 describes the used subtypes.

Message	Description
Authentication Request (*REQUEST*)	Authentication request message of a peer that contains its credentials. The Authentication Manager queries platform components to verify the peer's credentials. After successful verification, a security token is created and sent to the peer
Authentication Response (*REPLY*)	Response message to an authentication request message from SMARTCOM to the peer. It contains a security token that can be used in further requests to verify the identity of peer
Authentication Failed (*FAILED*)	Special response to an authentication request message from SMARTCOM to the peer that indicates that the authentication failed. The purpose of this message is to distinguish between the cases of a failed authentication and an authentication error on the basis of the message's subtype
Authentication Error (*ERROR*)	Special response message to an authentication message from SMARTCOM to the peer that indicates that there was an error during the authentication of the peer. Such an error might be that, for example, no external platform component is available that can verify the credentials

Table 3: Authentication Messages

7.2.3 *Message Adapters*

In order to use a specific communication channel, an associated *adapter* needs to be instantiated. Communication between peers and adapters is unidirectional — *output adapters* are used to send messages to peers; *input adapters* are used to receive messages from peers. SMARTCOM natively provides some common input/output adapters (e.g., SMTP/POP, Dropbox, Twitter). The role of adapters should be considered from functional and technical perspectives.

Functionally, the adapters allow for:

1. Hybridity – by enabling different communication channels to and from peers;
2. Scalability – by enabling SMARTCOM to cater to a dynamically changing number of peers;
3. Extensibility – new types of communication and collaboration channels can easily be added at a later stage transparently to the middleware's users.
4. Usability – human peers are not forced to use dedicated applications for collaboration, but rather freely communicate by relying on familiar third-party tools.
5. Load Reduction and Resilience – by requiring that all the feedback goes exclusively and unidirectionally through external tools first, only to be channelled/filtered later through a dedicated input adapter, SMARTCOM is effectively shielded from unwanted traffic load, delegating the initial traffic impact to the infrastructure of the external tools. At the same time, failure of a single adapter will not affect the overall functioning of the middleware.

Technically, the primary role of adapters is to perform the message format transformation. Optional functionalities include: message filtering, aggregation, encryption, acknowledging and delayed delivery. Similarly, the adapters are used to interface SMARTCOM with external software services, allowing virtualization of third-party tools as common software peers. The *Adapter Manager* is the component responsible for managing the adapter lifecycle (i.e., creation, execution and deletion of instances), elastically adjusting the number of active instances from a pool of available adapters. This allows the number of active adapter instances to be scaled as needed. This is especially important when dealing with human peers, due to their inherent periodicity, frequent instability and unavailability, as well as for managing a large number of connected devices, such as sensors. The Adapter Manager consists of the following subcomponents:

- *Adapter Handler*: managing adapter instance lifecycle. It handles the following adapter types:

  1. Stateful output adapters – output adapters that maintain conversation state (e.g., login information). For each peer a new instance of the adapter will be created;
  2. Stateless output adapters – output adapters that maintain no state. An instance of an adapter can send messages to multiple peers;
  3. Input pull adapters – adapters that actively poll software peers for feedback. They are created on demand by applications running on the platform and will check regularly for feedback on a given communication channel (e.g., check whether a file is present on an FTP server);
  4. Input push adapters – adapters that wait for feedback from peers.

- *Adapter Execution Engine*: executing the active adapters.
- *Address Resolver*: mapping adapter instances to peers' external identifiers (e.g., Skype/Twitter username) in order to initiate communication.

Input messages from peers (e.g., subtask results) or external tools (e.g., Dropbox file added, email received on a mailing list) are consumed by the adapters either by a push notification or by pulling at regular intervals (more details in Section 7.2.5). The principal adapter-handling algorithms are described in [191, 192].

7.2.4 Privacy Functionalities

SMARTCOM supports specification and observation of delivery and privacy policies on message, peer and collective levels:

- *Delivery policies* stipulate how to interpret and react to possible communication exceptions, such as: failed, timed out, unacknowledged or repeated delivery.
- *Privacy policies* restrict sending or receiving messages or private data to/from other peers, collectives or external applications under different circumstances.

Apart from offering predefined policies, SMARTCOM also allows users to import custom application- or peer-specific policies. As noted, both types of policies can be specified at different levels. For example, a peer may specify that he can be reached only by peer "manager" via communication channel "email", from 9a.m. to 5p.m. in collective "Work". The same person can set to be reachable via "SMS" any time by all collective members except "manager" in collective "Bowling". Similarly, a collective delivery policy may state that when sending instructions to a collective it suffices that the delivery to a single member succeeds to consider the overall delivery successful on the collective level. SMARTCOM takes care of combining and enforcing these policies transparently in different collective contexts.

7.2.5 Implementation & Evaluation

The SMARTCOM prototype was implemented in the Java programming language. One can interact with it through a set of provided APIs. The prototype comes with some implemented standard adapters (e.g., Email, Twitter, Dropbox), which can be used to test, evaluate and operate the system. Additional third-party adapters can be loaded as plug-ins and instantiated when needed. SMARTCOM uses MongoDB[8] as a database system for its various subsystems. Depending on the usage of the middleware, either an in-memory or dedicated database instances of MongoDB can be used. To decouple execution and communication we use Apache ActiveMQ[9] as the message broker. The source code is provided in SMARTCOM's GitHub repository[10].

As the envisioned positioning of SMARTCOM in the overall architecture of the incentive management platform requires that all information exchange takes place through it, the prototype was put through a performance evaluation to demonstrate that it is capable of withstanding high message loads (peaks) that might occur at specific times when an incentive may need be applied to a large group of workers simultaneously (e.g., a deadline).

The following performance evaluation was made on a 64-bit Intel Core2 Duo machine with 2x 2.53 GHz, 4.00 GB DDR2-RAM. The simulation configuration is as follows:

- One implementation of a Stateless Output Adapter (one instance shared by all peers).
- 10 Input Push Adapters to receive input from peers.
- Output and Input Adapters communicate directly using an in-memory queue to simulate a peer with a response time of zero.
- Worker threads ("Workers") simulate the concurrently executing incentive mechanisms sending incentive messages (rewarding actions) to "peers" (simulated human workers).
- One million messages are sent for each evaluation test run to get a meaningful average number of messages sent/received.

---

[8] http://www.mongodb.org
[9] http://activemq.apache.org
[10] https://github.com/tuwiendsg/SmartCom

- Only sent and received messages are considered as handled, not internal messages.

Figure 7.5 depicts the setup for the performance evaluation as described above.

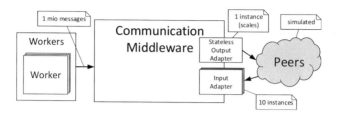

Fig. 7.5: Setup for the performance evaluations

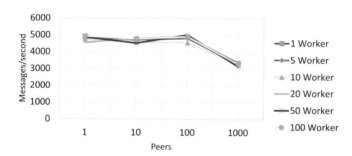

Fig. 7.6: Simulated message throughput. "Workers" are concurrent threads simulating
concurrent applications of rewarding actions to human "peers"

The performance was evaluated for every combination of 1, 5, 10, 20, 50 and
100 worker threads sending $1 \cdot 10^6$ messages concurrently, uniformly distributed to 1,
10, 100, and 1,000 peers waiting for messages and replying to them. Each test run
was executed 10 times to obtain average throughput results. Figure 7.6 presents the
results of the test runs. The test runs can be reproduced using the stated setup data
to configure the Java application located at GitHub[11]. As can be seen, the average
throughput remains between 5,000 and 3,000 messages per second. The performance
decrease with higher numbers of peers is the result of increased memory requirements
rather than computational complexity. The limiting factor here is the used ActiveMQ
message broker which only allows a maximum of approximately 20,000 messages
per second. The system has an upper bound of 5,000 messages per second since
each message is handled multiple times by the message broker and SMARTCOM.
This limitation applies to a single SMARTCOM instance, but multiple SMARTCOM

---

[11]   https://github.com/tuwiendsg/SmartCom/blob/master/smartcom-demo/src/
main/java/at/ac/tuwien/dsg/smartcom/demo/PerformanceDemo.java

instances can be deployed to balance the load if needed, sharing the database and peer store access. The chosen numbers of worker threads and peers cover the reasonably expected maximum numbers of concurrently executing incentive mechanisms and concurrently targeted humans, respectively. Performance (scalability) is, thus, not expected to become a primary concern of SMARTCOM, especially considering the inherent latency of human peers and variance of response times, which are both much higher in real-world than in simulated conditions.

7.3 Programming Model

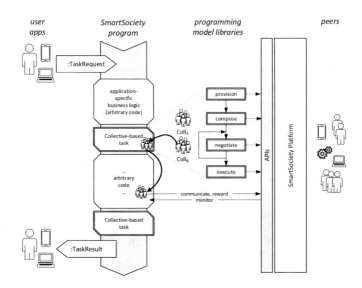

Fig. 7.7: Using the SmartSociety programming model

Figure 7.7 illustrates the intended usage of the programming model. The developer writes a SmartSociety program performing arbitrary business logic and handling the interaction with user applications. When a task requiring collaborative hybrid processing is needed, the developer uses the programming model library constructs to create and concurrently execute a *Collective-based Task (CBT)* – an object encapsulating all the necessary logic for managing complex collective-related operations: team provisioning and assembly, execution plan composition, human participation negotiations and finally the execution itself. These operations are provided by various SmartSociety platform components, which expose a set of APIs used by the programming model libraries. During the lifetime of a CBT, various *Collectives* related to the CBT are created and exposed to the developer for further (arbitrary) use in the remainder of the code, even outside of the context of the originating CBT or its lifespan. This allows the developer to communicate directly with the collective members, monitor and incentivize them, but also to use existing collectives to produce new ones, persist them, and pass them as inputs to other CBTs at a later point. In the remainder of the section, we will look in more detail into the design and functionality offered by CBT and Collective constructs.

7.3.1 Collective-Based Tasks (CBT)

A collective-based task (CBT) is the element of the programming model keeping the state and managing the lifecycle of a collective task. A CBT can be processed in one of the two collaboration models (*on demand* and *open call*) or a combination of the two, as specified by the developer. Table 4 lists the allowed combinations and describes them in more detail.

on_demand	open_call	allowed	description
true	true	yes	A collective of possible peers is first provisioned, then a set of possible execution plans is generated. The peers are then asked to negotiate on them, ultimately accepting one or failing (and possibly re-trying). The set of peers to execute the plan is a subset of the provisioned collective but established only at runtime. No known systems support it
true	false	yes	The expectedly optimal collective is provisioned, and given the task to execute. The task execution plan is implicitly assumed, or known before runtime. Therefore no composition is performed. Negotiation is trivial: accepting or rejecting the task. Example: Social Compute Unit (SCU) [25]
false	true	yes	"Continuous orchestration". No platform-driven provisioning takes place. The entire orchestration is fully peer driven (by arbitrarily distributed arrival of peer/user requests). The platform only manages and coordinates this process. Therefore, neither the composition of the collective nor the execution plan can be known in advance or vary over time, until either the final (binding) agreement is made, or the orchestration permanently fails due to non-fulfillment of some critical constraint (e.g., timeout). Note that in this case repetition of the process makes no sense, as the process lasts until either success or ultimate cancellation/failure (e.g., the ride request for traveling home for Christmas makes no sense after Christmas, or the user withdraws the request). Examples: pure ride-sharing, conventional crowdsourcing when negotiation is trivial
false	false	no	n/a

Table 4: CBT collaboration models and associated flags

At a CBT's core is a state machine (Fig. 7.8) managing transitions between states representing the eponymous phases of the task lifecycle: `provisioning`, `composition`, `negotiation` and `execution`. An additional state, named `continuous_orchestration`, is used to represent a state combining composition and negotiation under specific conditions, as explained in Table 4.

The purpose of the states is twofold: *a)* As in every state machine, to trigger certain actions upon transition into the state; *b)* to define allowed developer actions (i.e., to restrict certain function invocations to certain states). The choice of collaboration model is made through corresponding boolean flags, which are used in state transition guards to skip certain states and activities associated with them.

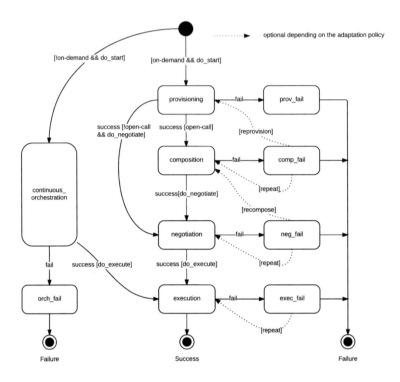

Fig. 7.8: CBT state diagram. States are associated with the task's lifecycle phases

7.3.1.1 provisioning State

On entering the `provisioning` state the provisioning process is started. The input collective equals a collective specified at CBT instantiation (most commonly a predefined collective representing all the peers from the Peer Manager eligible for that particular platform application). The provisioning process consists of running queries on the Peer Manager and provisioning algorithms (e.g., [25]) in search of appropriate human and software peers to participate in the subsequent execution, and setting up an infrastructural capability for their subsequent collaboration. Provisioning algorithms establish a set of human or machine peers that can support the computation, while being optimized on, e.g., highest aggregate set of skills, lowest aggregate price. The bootstrapping aspect refers to: finding and starting a software service, or inviting a human expert to sign up for the participation in the upcoming computation, and setting up the communication patterns among them (e.g., a shared Dropbox folder). The provisioning phase is crucial in supporting hybridity in the programming model, because it shifts the responsibility for explicitly specifying peer types or individual peers at design time from the developer onto the provisioning algorithms executed at runtime, thus making both human and machine-based peers eligible depending on the current availability of the peers and the developer-specified constraints. In practice, we assume that software peers are permanently available, and take into consideration only those human peers that are already registered to participate in a specific platform application. In this case, the provisioning process requires the following steps:

1. run queries in Peer Manager to find specific peers;
2. run a provisioning algorithm [25];
3. set up a communication topology [193] (e.g., set up a common communication medium, allow/prevent inter-/intra-collective communication).

7.3.1.2 composition State

The input to the state is the 'provisioned' collective. The composition is performed by the composition handler. The handler can be provided:

1. as a programming model library handler;
2. by the OM (ultimately by the dedicated OrchestrationManager (OM) instance).

In the composition process additional collectives are created and potentially exposed to the developer as a list of collectives named "negotiable", associated with composed execution plans. Upon success, the programming model waits for the flag *do_negotiate* to become true, and passes the composed execution plans and the associated "negotiable" collectives to the negotiation step. In case of failure, a dedicated failure state is entered. In case *b*), the composition state enacts the composition process.

7.3.1.3 negotiation State

The input to this state is the list of collectives named "negotiables" along with associated plans. The negotiation is performed by the negotiation handler. The handler can be provided:

1. as a programming model library handler;
2. by the OM (ultimately by the dedicated OrchestrationManager (OM) instance).

In case *b*), the negotiation state enacts the negotiation process. The outcome of the negotiation process is the single "agreed" collective and the associated execution plan.

Upon success, the programming model waits for the flag *do_execute* to become true, and passes the agreed execution plan and the associated "agreed" collective to the execution step. In case of failure, a dedicated failure state is entered.

7.3.1.4 continuous_orchestration State

In order to support continuous orchestration (as explained in Table 4) we need a state that does not separate composition and negotiation explicitly, but rather allows continuous switching between (re-)composing and negotiating, which allows users to submit new requests, which re-triggers composition, i.e., peers temporarily accept plans and later withdraw, until the plan is ultimately considered accepted and thus becomes ready for execution, or ultimately fails/gets canceled. Note that repetition of this state is not applicable, because repetition is generally done in case of remediable failures, but in this case the orchestration lasts until non-revocable success/failure.

7.3.1.5 execution State

The execution state handles the actual processing of the agreed execution plan by the "agreed" collective. In line with the general CAS principles, this process is intentionally made highly independent of the developer and the programming model and allowed be driven autonomously by the collective's member peers. Since peers can be either human or software agents, the execution may be either loosely orchestrated by human peer member(s), or executed as a traditional workflow, depending on what the state's handlers stipulate. For example, in the simplified collaborative software development scenario shown in Listing 7.4 both CBTs are executed by purely human-composed collectives. However, the testTask CBT could have been initialized with a different type, implying an execution handler using a software peer to execute a test suite on the software artifact previously produced by the progTask CBT. Whether the developer will choose software- or human-driven execution CBTs depends primarily on the nature of the task, but also on the expected execution duration, quality and reliability. In either case, the developer is limited to declaratively specifying the CBT's type, and the required termination criterion and the Quality of Results (QoR) expectations through associated handlers. The state is exited when the termination criterion evaluates to true. The outcome is "success" or "failure" based on the return value of the QoR handler. In either case, the developer can fetch the TaskResult object, containing the outcome, and the evaluation of the acceptability of the task's quality.

7.3.1.6 *_fail States

Each of the principal states has a dedicated failure state. Different failure states are introduced so that certain states can be re-entered, depending on what the selected adaptation policy specifies. Some failure states react only to specific adaptation policies; some to none. Adaptation policies are described in the following section.

7.3.2 Execution Model

A CBT instance is always associated with a `TaskRequest` containing input data and possibly a `TaskResult` containing the outcome of the task. Both are very generic interfaces meant to hide the application-specific format of the input and output data from the programming model, respectively. In fact, the programming model is designed to be *task-agnostic*. This is in line with the general CAS principle that unconstrained collaboration should be supported and preferred when possible. This design choice was made to allow subsequent support of different task description formats which will be interpretable by the application-specific orchestration manager (which is in charge of composing execution plans), or even by human peers only. This implies that the respective handlers (see below) registered for a specific application shall know how to interpret and produce correct formats of input and output data, and wrap them into `TaskRequest` and `TaskResult` objects.

The CBT has an independent execution thread that processes the initial task request by driving the associated state machine from initial until final state through required states by responding to events that allow state transitions and performing the associated workflows, and finally delivering the result. Each state has a predefined workflow that needs to be executed. The workflow consists of activities with predefined APIs. The activities act as placeholders for which different *handlers* can be registered. By registering different handlers for an activity, we can obtain different flavors, and ultimately a different overall execution of the state. For example, one of the activities in the Execution state is the "QoR" (quality of result). By being able to specify a different handler, we can produce different outcomes of the Execution phase. Similarly, by registering a different handler, an OM instance with different parameters can be used. This property is used to implement adaptation policies.

Figure 7.9 shows the workflows corresponding to the different states on the left, and the signatures of the delegates that can act as handlers for the corresponding activities. The programming model libraries provide a library of predetermined handlers, although external services from within the platform can also be used (e.g., [25]).

7.3.2.1 Adaptation Policies

An *adaptation policy* is used to enable re-doing of a particular subset of a CBT's general workflow with different functionality and parameters, by changing/re-attaching different/new handlers to the corresponding activities in the states' workflows, and enabling transitions from the failure states back to active states. The policies are triggered upon entering failure states, as shown in Figure 7.8. The possible transitions are marked with dotted lines in the state diagram, as certain policies make sense only in certain fail states.

Adaptation policies allow for completely changing the way a state is executed. For example, by registering a new handler for the `provisioning` state a different provisioning algorithm can be used. Similarly, a new handler installed by the adapta-

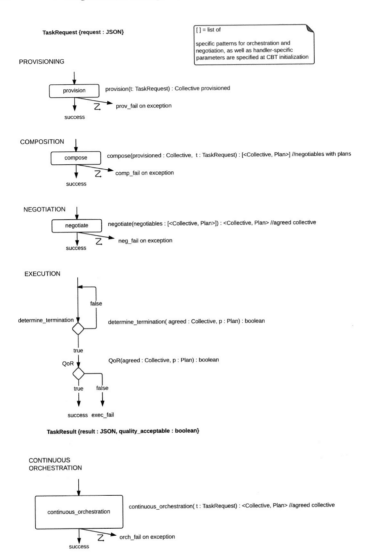

Fig. 7.9: CBT activity handlers, input and output types, and exceptions thrown

tion policy can in a repeated `negotiation` attempt use the "majority vote" pattern for reaching a decision, instead of the previous "consensus" pattern.

Since concrete adaptation policies are meant to extend the functionality of the programming model they are usually context-specific. Therefore, the programming model limits itself to offering the mechanism of extending the overall functionality through external policies and itself offers for each failure state only a limited set of simple, generally applicable *predefined policies*. In order to be general, predefined policies assume re-use of the existing handlers by default. The supported policies are informally described in Table 5. Only a single adaptation policy is applicable in a single failure state at a given time. If no policy is specified by the developer, the ABORT policy is assumed (shown as full-line transition).

adaptation policy	description
ABORT	Default. Do nothing, and let the fail state lead to total failure
REPEAT	Repeats the corresponding active state, with (optionally) new handler(s). If the developer specifies the new handler we describe the property as "adaptivity"; if the system automatically determines the new handler, we describe the property as "elasticity"
REPROVISION	Transition into provisioning state, with (optionally) a new provisioning handler
RECOMPOSE	Repeat the composition, with (optionally) a new provisioning handler

Table 5: CBT adaptation policies

7.3.3 Collectives

The notion of collective is very general. Sometimes it used to denote a stable group or category of peers based on their common properties, but not necessarily with any personal/professional relationships (e.g., "Java developers", "students", "Vienna residents", peers registered with a specific application); in other cases, the term collective was used to refer to a team – a group of people gathered around a concrete task. The former type of collectives is more durable, whereas the latter one is short-lived. Therefore, we make the following distinction in the programming model:

A **Resident Collective (RC)** is an entity defined by a persistent Peer Manager identifier, existing across multiple application executions, and possibly different applications. Resident collectives can also be created, altered and destroyed fully out of scope of the code managed by the programming model. The control of who can access and read a resident collective is enforced by the Peer Manager. For those resident collectives accessible from the given application, a developer can read/access individual collective members as well as all accessible attributes defined in the collective's profile. When accessing or creating an RC, the programming model either passes to the Peer Manager a query and constructs it from returned peers, or passes an ID to get an existing Peer Manager collective. In either case, in the background, the programming model will pass to the Peer Manager its credentials. So, it is up to the Peer Manager to decide based on the privacy rules which peers will get exposed (returned). For example, for "TrentoResidents" we may get all Trento residents who have nothing against participating in a new (our) application, but not necessarily all Trento residents from the Peer Manager's database. By default, the newly created RC remains visible to future runs of the application that created it, but not to other applications. The Peer Manager can make them visible to other applications as well. At least one RC must exist in the application, namely the collective representing all peers visible to the application.

Application-Based Collective (ABC). Differently from a resident collective, an ABC's lifecycle is managed exclusively by the SmartSociety application. Therefore, an ABC cannot be accessed outside of an application, and ceases to exist when the application's execution stops. ABCs are initially created by applying certain operations on resident collectives. Differently than resident collectives, an ABC is an atomic entity for the developer, meaning that individual peers cannot be explicitly known or accessed from an ABC instance. This means that an ABC instance is immutable. The ABC is the embodiment of the principle of collectiveness, making the collective an atomic, first-class citizen in our programming model. Concretely, the following rules regulate the ABCs' lifecycle:

- An ABC can be created:

    - implicitly, as an intermediate product of different states of CBT execution (e.g., "provisioned", "agreed").
    - explicitly, by using dedicated collective manipulation operators:
        · to clone a resident collective.
        · as a result of a set operation on two other Collectives (either RC or ABC).

- An ABC has a lifetime equal to that of the application where it was created:

    - the lifecycle of an ABC is managed by the `SmartSocietyApplication Context` (CTX) (Sect. 7.3.4)
    - the Developer cannot explicitly delete them; they are cleaned up by the runtime when the object representing the application is finalized.

- An ABC is:

    - immutable for the developer.
    - atomic for the developer (i.e., individual peers cannot be accessed from it).

The context object (Sect. 7.3.4) registers and keeps the collective's *kind*. The kind of a collective is a CTX record describing the allowed Peer Manager peer profiles for the given collective type, as well as the overall collective entity profile [29]. Furthermore, once full privacy policies are implemented, this will allow fine-grained control of which peer profiles and attributes are exposed to the application developer for a given collective kind, implemented between CTX and Peer Manager. The Peer Manager retains the ultimate right to grant/restrict access to specific peers or their attributes based on the privacy policies. So, in a way, the kind can be though of as a CTX-specific type of the collective. Therefore, ABCs are meaningful only within a context that describes their kind and the types of profiles they contain. By pre-registering the kind in the CTX, the runtime announces to the Peer Manager the wish to access and store specific attributes of the collective and its peers.

Figure 7.10 depicts a partial UML class diagram showing the relevant collective classes and part of the API. Concrete methods are explained in detail in Section 7.4.

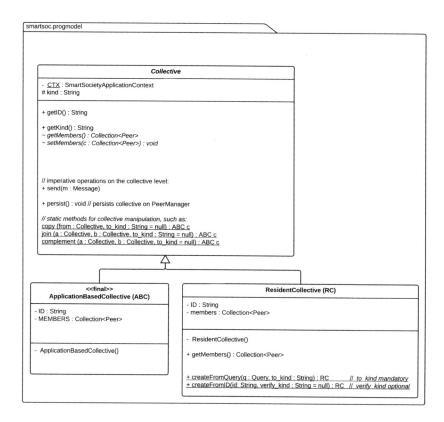

Fig. 7.10: Partial UML class diagram showing supported collective manipulation methods

7.3.4 Application Context & Initialization

The application context (CTX) represents a particular instance of the SmartSociety application (class SmartSocietyApplicationContext). It can be described in more detail with the following claims:

- There must be exactly one CTX instance per application run. Therefore, CTX is a singleton.
- The CTX instance keeps all values, overall application state and metadata initialized for the application run. Therefore, all initialization methods are over CTX.
- The CTX represents the application, therefore it possesses the internal business logic to track lifecycles of all created ABC instances, and other entities created for the purposes of the application.

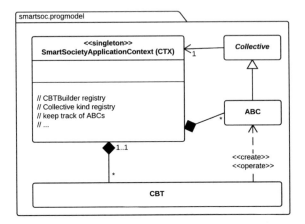

Fig. 7.11: Partial UML class diagram showing important relationships between CTX and other programming model elements

For the CTX initialization the developer is required to provide some information that will then be used during task request handling:

- internal profile: contains metadata associated with the application in the Peer Manager that can be required by clients or other applications;
- internal profile schema: the Peer Manager requires that a schema be associated with a profile;
- user profile schemas: peers registered to the application will be seen through a profile that is application-specific; the CTX will register such a schema with the Peer Manager;
- collective kind descriptors: collectives will be created according to some specification depending on their type and must abide by it; the CTX is in charge of providing the registry for the different kinds in order to be able to enforce such rules when required;
- CBT builders for CBT types: CBTs may behave differently according to their goal. For this reason for each CBT type the developer must provide a builder (CBTBuilder) that will take care to create the proper CBT when required.

Section 7.4.1 illustrates the language construct to provide all the required information.

Composition

The OM posts the TaskRequest that is received from the CBT, together with the provisioned collective, to the OM. Composition may or may not be trivial and outputs a list of pairs, where each pair is an execution plan and a negotiable collective for that plan. The OM performs the necessary functionality in order to determine the completion of the execution of composition by using the API that is defined for the OM. Once composition ends, the OM forms the list of pairs of negotiable collectives and the respective plans in which negotiation can occur for these collectives. This list of pairs is returned to the CBT and composition is now over at the level of the CBT that invoked the call on the OM for composition.

Negotiation

At the beginning of the negotiation phase, the OM uses SMARTCOM so that all the members of the negotiable collectives can be contacted and thus be notified that a new plan is waiting for them in which they can potentially agree or disagree to participate. The OM translates accept/reject messages that conform into the specification of the SMARTCOM middleware to the appropriate operations in the negotiation protocol. The response of the OM is translated into the appropriate SMARTCOM message and transmitted back to the user that attempted the current negotiation (accept/reject).

Execution

At the beginning of the execution phase, each member of the collective that has agreed to participate in the execution gets contacted via SMARTCOM is sent the agreed execution Plan. The subsequent execution is independent of the programming model, and is potentially monitored by other SmartSociety/Smart City platform components.

Continuous Orchestration

Composition runs for every new TaskRequest that arrives in the system. The outcome of composition is a set of plans where in each plan we have a separate collective where negotiation can be performed at a later stage. Composition is similar to the case where a provisioned collective is supplied. Composition is performed and a set of Plans associated with the TaskRequest that has just arrived are generated. However, as these plans are shared among different TaskRequests, it is usually the case that additional TaskRequests are affected that belong to other users (and for which composition has already run earlier). In any case, the negotiable collectives - one such collective per plan generated - are passed as parameters to the negotiation phase together with their plans (they are in 1-1 correspondence in the list of pairs that is the argument). During the composition phase, continued orchestration may also throw a CompositionFailedException, where depending on the adaptation policy this may or may not result in an overall failure for some associated CBTs.

Similarly to the case described earlier, at the beginning of the negotiation phase, and after potentially contacting the OM to initialize the negotiation, the SMARTCOM middleware contacts each collective and delivers the plan that was just generated through composition for the collective and on which negotiation may start.

7.4 Language Constructs

construct	operating on	declarative/imperative
collaboration pattern	CBT	D
CBT activity handler	CBT	D
adaptation policy	CBT	D
CBT lifecycle command	CBT	I
CBT collective-fetching	CBT	I
collective manipulation	Collective	I
communication	Collective	I
initialization constructs	CTX	D

Table 6: Supported language constructs and associated programming model elements

Table 6 summarizes language constructs supported by the programming model and lists the elements of the model on which they operate, and whether they are declarative or imperative. The imperative constructs for CBT are used to control entry into different states and other CBT lifecycle operations (Sect. 7.4.3). The imperative constructs for collectives are those meant for fetching, creating, merging and copying collectives, reading attributes, and exchanging messages with collectives (Sect. 7.4.5) The remaining constructs are declarative. They need to be specified before the element of the model that they describe gets instantiated.

7.4.1 Context Initialization Constructs

During the CTX initialization the developer provides the required information that will be used during the application execution (Section 7.3.4). The programming model allows the developer to submit such information through the following methods:

- void updateInternaProfile(ProfileSchema profileSchema,
  Profile profile):
  the internal profile is created and updated, in case of creation the profileSchema is also provided to the Peer Manager;
- void registerUserProfileSchema(ProfileSchema profileSchema):
  the schema of the profile that will be used to store user data;
- void collectiveForKind(String kind,
  CollectiveDescriptor descriptor):
  for each collective kind a corresponding descriptor is registered;

- void registerBuilderForCBTType(String cbtType,
  CBTBuilder builder):
  this method specifies how CBT of a certain type can be built. Several CBTBuilder
  implementations can be provided as part of the programming model library; in
  particular there is one that makes use of the external Orchestration Manager.

In this section some examples of application context initialization are given.

7.4.2 CBT Instantiation

The CBT is instantiated through a CBTBuilder. In order to offer a human-friendly
and comprehensible syntax in conditions where many parameters need to be passed
at once, we make use of the nested builder pattern to create a "fluent interface"[12] for
CBT instantiation, as exemplified in Listing 7.2.

Builders are registered at CTX initialization time and associated with a CBT type
(Section 7.4.1). Once the right type builder is retrieved all the expected parameters
must be passed to the builder (according to its actual type), and then the build must
be called (as shown in Listing 7.2).

```
1  /* ... */
2  CBT cbt = ctx.getCBTBuilder("RideSharingType")
3                .of(CollaborationType.OC) //Enum: OC, OD, OC_OD
4                .forInputCollective(c)
5                .forTaskRequest(t)
6                .withNegotiationArgs(myNegotiationArgs)
7                .build();
8  /* ... */
```

Listing 7.2: Instantiatiation of a CBT

7.4.3 CBT Lifecycle Operations

The basic method for checking the lifecycle state is the following. It returns an
enumeration CBTState with the active state of the CBT.

- CBTState getCurrentState() – Returns the current state.

We additionally provide the following utility (pretty-print) methods for comparing
the state of execution:

---

[12] http://www.martinfowler.com/bliki/FluentInterface.html

- boolean isAfter(CBTState compareWith) – returns true if the CBT has finished executing the compareWith state; this also includes waiting on the subsequent state. Throws exception if the comparison is illogical.
- boolean isBefore(CBTState compareWith) – returns true if the CBT has not yet started executing the compareWith state; this also includes waiting for the compareWith state. Throws exception if the comparison is illogical.
- boolean isIn(CBTState compareWith) – checks whether compareWith is equal to the return value of getCurrentState().

The do_statename flags (Fig. 7.8) are used in guard conditions that control CBT's state transitions. Apart from the main CBT states shown in Fig. 7.8, between each two states there is an intermediate state, named waiting_for_nextstate, in order to allow blocking until the corresponding guard condition is met. If a flag for transitioning further is not enabled, the CBT will remain (blocked) in the current intermediate state until the flag is set, or transition to the fail state of the originating main state after a timeout is exceeded. Upon instantiating a CBT, the developer defines whether the state transition should happen automatically, or be explicitly controlled. In order to check for these states, we expose the following set of methods:

- boolean isWaitingForProvisioning()
- boolean isWaitingForComposition()
- boolean isWaitingForNegotiation()
- boolean isWaitingForContinuousOrchestration()
- boolean isWaitingForStart() – waiting in the initial state to enter any main state.

Furthermore, we have the following related methods:

- boolean isRunning() – – true in every state except initial or final.
- boolean isDone() – true only in the final state (either success or fail), not matter whether we arrived in it through success or one of the fail states.

To allow the developer to control CBT transitions explicitly the developer is offered the following constructs to get/set the flags used in guard conditions and wake up the CBT's thread if it was waiting on this flag:

- get/setDoCompose(boolean tf)
- get/setDoNegotiate(boolean tf)
- get/setDoExecute(boolean tf)

By default, the CBT gets instantiated with all flags set to true. We also provide a convenient method that will simultaneously set all flags to true/false:

- setAllTransitionsTo(boolean tf)

Since from the initial state we can transition into more than one state, for that we use the method:

- `void start()` – Non-blocking call. Allows entry into the `provisioning` or `continuous_orchestration` states, depending on which of them is the first state.

Additionally, CBT exposes a number of additional methods to match the methods offered by the Java 7 `Future` API:

- `TaskResult get()` – waits if necessary for the computation to complete (until `isDone() == true`), and then retrieves its result. Blocking call.
- `TaskResult get(long timeout, TimeUnit unit)` – same as above, but throwing appropriate exception if timeout expired before the result was obtained.
- `boolean cancel(boolean mayInterruptIfRunning)` – attempts to abort the overall execution in any state and transition directly to the final fail state. The original Java 7 semantics of the method is preserved.
- `boolean isCancelled()` – Returns true if CBT was canceled before it completed. The original Java 7 semantics of the method is preserved.

7.4.4 CBT Collective-Fetching Operations

A CBT object exposes the following methods for fetching the ABCs created during the CBT's lifecycle:

- `Collective getCollectiveInput()` – returns the collective that was used as the input for the CBT.
- `ABC getCollectiveProvisioned()` – returns the "provisioned" collective
- `ABC getCollectiveAgreed()` – returns the "agreed" collective.
- `List<ABC> getNegotiables()` – returns the list of negotiable collectives.

At the beginning of a CBT's lifecycle, the return values of these methods are *null*/empty list. During the execution of the CBT, the executing thread updates them with current values. Note that we never return `Plans`, as the programming model is task-agnostic.

7.4.5 Collective Manipulation Constructs

As noted in Section 7.3.3 resident collectives (RCs) are created by querying the Peer Manager via the following static methods of the `ResidentCollective` class:

- `ResidentCollective createFromQuery(PeerMgrQuery q, string to_kind)` – Creates and registers a collective with the Peer Manager. It is assumed that the kind entity describing the new collective has been properly registered and initialized with CTX. When contacting the Peer Manager we pass also the ID of the application, and we assume that the Peer Manager checks (with the help of the programming model runtime) whether we are allowed to create a collective of the requested kind, and returns only those peers whose privacy settings allow them to be visible to our application's queries. The registered kind descriptor in the CTX allows this method to know how to properly transform the attributes from the entities obtained from the Peer Manager to those expected by the target kind.

- `ResidentCollective createFromID(string ID, string to_kind)` – Creates a local representation of an already existing collective on the Peer Manager, with a pre-existing ID. Invocation of this method will not create a collective on the Peer Manager, so in case of passing a non-existing collective ID an exception is thrown. This method allows us to use and access externally defined RCs. When contacting the Peer Manager we pass also the ID of the application, and we assume that the Peer Manager checks whether we are allowed to access the requested collective, and returns only those peers whose privacy settings allow them to be visible to our application's queries. The registered kind descriptor in the CTX allows this method to know how to properly transform the attributes from the entities obtained from the Peer Manager into those expected by the target kind.

On the other hand, ABCs are created from existing collectives (both RCs and ABCs) through the following static methods of the `Collective` class:

- `ABC copy(Collective from, [string to_kind])` – Creates an ABC instance of kind `to_kind`. Peers from collective `from` are copied to the returned ABC instance. If `to_kind` is omitted, the kind of collective `from` is assumed.
- `ABC join(Collective master, Collective slave,` `[string to_kind])` – Creates an ABC instance containing the union of peers from collectives `master` and `slave`. The resulting collective must be transformable into `to_kind`. The last argument can be omitted if both `master` and `slave` have the same kind.
- `ABC complement(Collective master, Collective slave,` `[string to_kind])` – Creates an ABC instance containing the peers from collective master after removing the peers present both in master and in slave. The resulting collective must be transformable into `to_kind`. The last argument can be omitted if both `master` and `slave` have the same kind.
- `void persist()` – Persist the collective on Peer Manager. RCs are already persisted, so in this case the operation defaults to renaming. In case of an ABC, the Peer Manager persists the collective as an RC. However, this does not mean that the developer is able to subsequently fetch that RC and access the collective members. This is decided by the CTX and Peer Manager based on the ABC's kind.

To read the ABC's attributes, the following `ABC` class method is used:

- `Attribute getAttribute()` – searches attribute fields then returns a clone of the found `Attribute`, if any present.

7.4.6 Collective-Level Communication

Programming model libraries fully rely on SMARTCOM to support the communication with collectives. At the moment, the programming model only allows a basic set of communication constructs, namely those for sending a message to a hybrid collective and receiving responses from it:

- `void send(Message m) throws CommunicationException` – Send a message to the collective. Non-blocking. Does not wait for the sending to succeed or fail. Errors and exceptions thereafter will be sent to the Notification Callback.
- `Identifier registerNotificationCallback(NotificationCallback onReceive)` – Register a notification callback method that will be called when new messages from the collective are received. The returned `Identifier` is used for unsubscribing.
- `void unregisterNotificationCallback(Identifier callback)` – unregister a previously registered callback.

These methods are invokable on any `Collective` object. The delivery is in line with individual privacy preferences. The types `Message`, `Identifier` and `NotificationCallback` are described in [192]. In the background, the programming model hides from the developer the complexity of invoking SMARTCOM's `Communication` API directly, and the need to register output adapters. Furthermore, if the `provisioning` state is used, additional virtual communication infrastructure (i.e., output adapters) may be set up transparently to the developer.

7.4.7 Examples

Initializing a CBT

```
1  void ctx_initialization() {
2    /* Registering OM based builder for CBT of type "RideSharing" */
3    MyOM OM = new MyOM();
4    CBTBuilder builderWithOM = new OMBasedCBTBuilder(OM);
5    ctx.registerBuilderForCBTType("RideSharing", builderWithOM);
6
7    /* Registering another CBT type for which library provided handlers
8       are used. */
9    CBTBuilder patternBasedCBTBuilder = new PatternBasedCBTBuilder();
10   ctx.registerBuilderForCBTType("SimpleTask", patternBasedCBTBuilder);
11 }
12
13 /* this method is called for requests requiring a "RideSharing" CBT */
14 CBT get_CBT_for_rideSharing(TaskRequest request) {
15   CBT cbt =
16     ctx.getCBTBuilder("RideSharing")
17       .of(CollaborationType.OC)
18       .forInputCollective(c) //c is some input collective
19       .forTaskRequest(request)
20       .withNegotiationArguments(5, TimeUnit.Days)
21       .build();
22 }
23
24 /* this method is called for requests involving a "SimpleTask" request */
25 CBT get_CBT_for_rideSharing(TaskRequest request) {
26   CBT cbt =
27     ctx.getCBTBuilder("SimpleTask")
28       .of(CollaborationType.OC)
29       .forTaskRequest(request)
30       .withCompositionArguments(CompositionPattern.PROVIDED_COLLECTIVE)
31       .withNegotiationArguments(
32                      NegotiationPattern.AGREEMENT_RELATIVE_THRESHOLD,0.6)
33       .build();
34 }
```

Listing 7.3: CBT initialization

Listing 7.3 shows an example of how a CBT can be initialized. The pieces of code are grouped in dummy methods for presentation purposes.

The CTX initialization phase (lines 1-11) is executed after the CTX has been created. The developer must provide a CBTBuilder for each type of CBT that is going to be used during the application lifetime. Two different types are registered, RideSharing and SimpleTask; for each of them a different builder is provided. For the sake of usability, a support library provided along with the programming model will provide several CBTBuilder implementations.

For the type RideSharing the developer decides to use an OM, the OM is instantiated in advance (the CTX is not involved), and then it is used to create a generic builder based on OM (a library-provided one). Also for the SimpleTask type the developer makes use of a given builder. Comparing the CBT creation code for the two types (lines 14-22 for the RideSharing type, lines 25-34 for

SimpleTask), one can appreciate the consistency of the CBTBuilder interface. However the different builder implementations will require different arguments for the creation of handlers. For instance in the example the OM-based builder requires a timeout for the negotiation and the user can create it. Note how for some handlers no parameter is passed; this might means that the builder does not allow any parametrization in the handler creation, or it might provide a default behavior. In the second example the developer does not provide an input collection, it specifies instead the behavior both for composition and negotiation.

Manipulating and reusing collectives

Consider an application that uses the SmartSociety platform to assemble ad hoc, on-demand programming teams to build software artifacts. For this purpose, two CBT types are registered with CTX: "MyJavaProgrammingTask" and "MyJavaT-estingTask". First, the developer creates an RC javaDevs containing all accessible Java developers from the Peer Manager. This collective is used as the input collective of the progTask CBT. progTask is instantiated as an on-demand collective task, meaning that the composition state will be skipped, since the execution plan is implied from the task request myImplementationTReq.

The collective is first processed in the provisioning phase, where a subset of programmers with particular skills are selected and a joint code repository is set for them to use. The output of the provisioning state is the "provisioned" collective, a CBT-built ABC collective, containing the selected programmers. Since it is atomic and immutable, the exact programmers which are members of the team are not known to the application developer. The rationale here is similar to cloud computing – the user specifies the infrastructural requirements and constraints and the platform takes care to provision this infrastructure, without letting the user care about which particular VM instances are used.

The negotiation pattern will select the first 50% of the provisioned developers into the 'agreed' collective that will ultimately execute the programming task. After the progTask's this ABC becomes exposed to the developer, which uses it to construct another collective, containing Java developers from the 'provisioned' collective that were not selected for the "agreed" one. This collective is then used to perform the second CBT testTask, which takes as input the output of the first CBT.

```
 2  Collective javaDevs = ResidentCollective.createFromQuery(myQry,"JAVA_DEVS");

 4  CBT progTask = ctx.getCBTBuilder("MyJavaProgrammingTask")
 5                     .of(CollaborationType.OD)
 6                     .forInputCollective(javaDevs)
 7                     .forTaskRequest(myImplementationTReq)
 8                     .withNegotiationArguments(
 9                         NegotiationPattern.AGREEMENT_RELATIVE_THRESHOLD, 0.5)
10                     .build();

12  progTask.start();

14  //    ...
15  //    assume negotiation on progTask done
16  //    ...

18  Collective testTeam; //will be ABC
19  if (progTask.isAfter(CBTState.NEGOTIATION)) {
20    // out of provisioned developers, use the other half for testing
21    testTeam = Collective.complement(
22        progTask.getCollectiveProvisioned(), progTask.getCollectiveAgreed());
23  }

25  while (!progTask.isDone()) { /* do stuff or block on condition */}

27  TaskResult progTRes = progTask.get();

29  if (! progTRes.isQoRGoodEnough()) return;

31  CBT testTask = ctx.getCBTBuilder("MyJavaTestingTask")
32                     .of(CollaborationType.OD)
33                     .forInputCollective(javaDevs)
34                     .forTaskRequest(new TaskRequest(progTRes))
35                     .withNegotiationArguments(
36                         NegotiationPattern.AGREEMENT_RELATIVE_THRESHOLD, 1.0)
37                     .build();
38  /*...*/
```

Listing 7.4: Using the "agreed" and "provisioned" ABCs to obtain a third collective that will be used in another task. Also, using the outcome of one CBT in another one

Controlling CBT execution

The following code snippet shows some examples of interaction with a CBT lifecycle. An on-demand CBT named cbt is initially instantiated. For illustration purposes we make sure that all the transition flags are enabled (true by default), then manually set do_negotiate to false, to force cbt to block before entering the negotiation state, and start the CBT. While the CBT is executing, arbitrary business logic can be performed in parallel. At some point, the CBT is ready to start negotiations. At that moment, for the sake of the demonstration, we dispatch the motivating messages (or possibly other incentive mechanisms) to the human members of the collective, and let the negotiation process begin. Finally, we block the main thread of the application waiting on the `cbt` to finish or the specified timeout to elapse, in which case we explicitly cancel the execution.

```
2  CBT cbt =  /*... assume on_demand = true ... */

4  cbt.setAllTransitionsTo(true); //optional
5  cbt.setDoNegotiate(false);
6  cbt.start();

8  while (cbt.isRunning() && !cbt.isWaitingForNegotiation()) {
9      //do stuff...
10 }

12 for (ABC negotiatingCol : cbt.getNegotiables() {
13     negotiatingCol.send(new SmartCom.Message("Please accept this task"));
14     // negotiatingCol.applyIncentive("SOME_INCENTIVE_ID");
15 }
16 cbt.setDoNegotiate(true);

18 TaskResult result = null;

20 try {
21     result = cbt.get(5, TimeUnit.HOURS); //Blocks until done, but max 5 h
22     // do something with result
23 }catch(TimeoutException ex) {
24     if (cbt.getCollectiveAgreed() != null){
25         cbt.getCollectiveAgreed().send(
26                 new SmartCom.Message("Thanks anyway, but too late."));
27     }
28     cbt.cancel(true);
29 }

31 //...
```

Listing 7.5: Top: Incentivizing a collective about to start negotiating. Bottom: Canceling a task taking too long

Chapter 8
Incentive Management

Most state-of-the-art socio-technical platforms model humans as computing elements or services with predefined functionalities (see Section 6.1). While this is an abstraction suitable for efficient algorithmic querying, composition and execution planning, it treats the human participant as a computing node with statistical probabilities with respect to properties such as availability, response time, effort level and performance quality. However, due to the high unpredictability of human nature and a multitude of unforeseeable external factors, as well as difficulties and costs associated with measuring these metrics, this abstraction is often inaccurate for practical purposes. Even more importantly, existing platforms limit themselves to trying to model and measure these properties without attempting to actively influence them. As we will show in this chapter, actively motivating and engaging humans (citizens) can bring many benefits compared to existing approaches, in terms of participation willingness, increased quality, performance and collectiveness, but most importantly in eliciting the creative and cognitive potential of the human participants.

Incentives are a well-established method for influencing and motivating humans and for aligning otherwise diverging individual interests towards a common target. They are researched and used in all scientific disciplines concerned with organization of work and collective activities, extending far beyond the usually implied monetary rewards. Incentives gain importance particularly in Smart City environments, where attracting and organizing the citizens to participate in joint collaborative activities is difficult due to the number and diversity of potential participants and the scale and complexity of collaborations. Furthermore, they can be useful to the citizens for comparing similar Smart City services and for transferring their own reputation among those services by showcasing the obtained rewards.

In this chapter we introduce the field of automated incentive management and present our research in the area. We build upon the theoretical basis to design and present a complete methodology and a software framework prototype for automated incentive management in socio-technical systems applicable to Smart City environments.

© Springer International Publishing AG 2017
S. Dustdar et al., *Smart Cities*,
DOI 10.1007/978-3-319-60030-7_8

8.1 Research Context

Many look upon social computing and crowdsourcing with criticism. The main argument of such criticisms is that it is not ethical to treat humans as computing nodes, and simply exploit them on a pay-as-you-go basis, which in practice often means engaging them for microtasks that earn sub-dollar amounts. We can observe striking similarities between the evolution of worker rights in conventional industry and in the digital domain. Workers in early industry were exploited by working in unregulated working environments, working on repetitive activities for long hours. Yet, at the time, for many workers this was the best choice. Today's crowdsourcing landscape is still very much unregulated, and can be perceived as exploitative. However, studies such as [107] show that for a number of people such work represents a secure and important source of income. Recognizing the importance of digital labor, there has been a push from unions and crowd workers themselves to regulate digital markets. Platforms such as Turkopticon [81] allow workers to identify unfair employers, raise awareness of appropriate compensation and offer a social environment to discuss related problems and seek advice. The largest German worker union IG Metall has started an initiative[1] involving industry and academia to monitor and regulate the digital labor markets. Such efforts represent important initial steps in improving the working conditions in the digital market, and raising awareness of the importance of it. A further step in this direction is to make such markets attractive to highly skilled labor, and making them easily accessible to the general public. This will increase the diversity in types and complexity of performed jobs but also allow the markets to get better regulated, both through self-regulation due to the highly dynamic relationship between supply and demand of jobs, as well as through the increased interest of unions and governmental agencies.

A 2016 global report[2] by the World Economic Forum on the future of jobs shows that the changing nature of work is expected to cause the biggest impact on global labor markets, fueled by technological drivers such as cloud computing, IoT, big data, crowdsourcing and the sharing economy. Companies are expected to employ a smaller permanent (core) workforce, and to scale out when additional expertise or processing power is needed. The same report shows that the global ease of recruitment is already low in all economic sectors, and the trend is forecast to significantly worsen in the near future. Such prospects put the integrative socio-technical vision of the Cyber-Human Smart City at the center of attention, as it represents a natural market for offering and performing complex, cognitive and physical collaborative tasks drawing on to workers from the general population. Innovative companies can leverage this great potential in human resources and make parts of their businesses much more dynamic and scalable (Figure 8.1).

---

[1] http://www.faircrowdwork.org/en

[2] Survey of 371 companies representing more than 13 million employees across nine broad industry sectors in 15 major developed and emerging economies and regional economic areas: https://www.weforum.org/reports/the-future-of-jobs

However, such dynamic working environments are especially attractive to exploitative activities due to the existing diversity of working roles, the scale of workforce, the transient nature of the employment relationship and complexity of tasks performed collectively. In such environments, incentives are being increasingly used to prevent the various types of dysfunctional behavior occurring.

traditional company   crowdsourcing company   "socially-enhanced computing" company

Fig. 8.1: Evolution of digital labor: As working patterns become more complex, social computing platforms need advanced organizational structure and "crowd" management capabilities, including automated incentive management

Even without the added complexity characteristic of socio-technical environments, incentives play a major role in managing large-scale workforces. This is evidenced by the fact that most big or medium-sized companies employ some incentive measures; e.g., over 80% in the USA [51, Chap. 1]. Furthermore, numerous studies have shown the effectiveness [136] of different incentive mechanisms and their selective and motivating effects [98]. In crowdsourcing environments, and consequently in Cyber-Human Smart Cities, incentives are expected to play an even more important role. In the visionary paper [89], a number of leading authors in the area discuss prospectives of crowdsourcing and identify *incentive management* as one of the key research directions. However, contemporary approaches to incentive management usually imply hard-coded, system-specific solutions. Such approaches are not portable, and prevent reuse of common incentive logic. That hinders cross-platform application of incentives and reputation transfer. Additionally, in future Smart City environments there will be a need to combine, personalize and frequently adapt incentive mechanisms [178], something that is not straightforward with current approaches. An additional challenge is that modeling of incentives is performed by multidisciplinary domain experts (*incentive designers*) often lacking knowledge of the technical internals of the socio-technical platform. On the other hand, the platform developers lack the domain knowledge necessary to understand the provided incentives, leading to a discrepancy between modeling and implementation processes.

Figure 8.2 visualizes the application context of an *incentive management framework*: A complex collaborative process is being executed by employing crowdsourced team(s) of human experts to execute various collective activities. The teams are provisioned by the Smart City platform component that assembles teams based on required functionality, collaboration patterns and elasticity parameters such as: price, speed

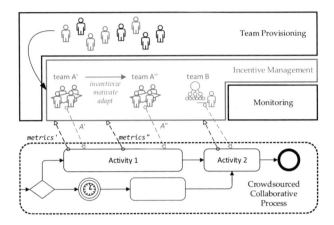

Fig. 8.2: Application context of incentive management systems

or reputation. In order to monitor and influence the behavior of the team's members during and across activity executions an incentive scheme needs to be enacted.

This is the task of the envisioned incentive management framework. It enacts the incentive scheme by applying rewards or penalties in a timely manner to induce wanted worker behavior, thus effectively performing runtime team adaptations (see Fig. 8.2: $A' \rightarrow A''$). Most real-world incentive strategies can be composed of modelable and reusable bits of incentive logic ([155, 172]). However, the efficacy of incentives can depend on multiple other factors, such as team size, cultural background or knowledge of other participants ([51]). The scheme is usually a result of a prior assessment or case study of the particular application scenario, but needs subsequent adaptations and adjustments [178]. Therefore, the challenge is to design an incentive management framework capable of combining and reusing existing and proven incentive mechanisms, but also allowing for easy tweaking to particular application contexts. We make a strong case for this in Section 8.2.4.3

Prior to enactment, an incentive scheme must be modeled and encoded by an incentive designer. The *incentive model* used in the process needs to be based on widely adopted incentive practices in both traditional companies and in contemporary social computing environments to allow for expression of realistic incentives covering a wide array of incentivizing use cases.

A Word About Terminology

Research on incentives originates from economics, where the incentives were first used as a means of aligning the interests of blue-collar workers with those of the business owners/employers. Even though our understanding and perception of incentives has evolved to appreciate the mutual benefits of incentives to both parties and has expanded across all possible types of cognitive and physical work, the originating theories are still very much reminiscent of the traditional production-line manufacturing environments with respect to the terminology used. This is why we commonly encounter terms such as *principal/owner/employer* labeling the providers of the incentives, and the term *worker* denoting the receiver/consumer of the incentive without any further assumptions about the nature of their relationship. As these have become commonly used terms in the literature, we also use them throughout this chapter.

However, with the advent of social computing, researchers from other areas started investigating incentives borrowing the terms typical of those areas and introducing them into incentives research. This is why we can often find the term *agent/peer* denoting the consumer of the incentive in a socio-technical environment. The appearance of digital labor also changed the perception of the working process itself and the working environment. We now talk of *tasks/jobs/workflows* offered by a *platform* acting as an intermediary between the actual task (and incentive) provider and the humans performing the work and consuming the incentive. A Smart City research environment is inevitably a multidisciplinary one. In the given situation we use all the outlined terms interchangeably where they feel more natural.

8.2 Existing Incentive and Rewarding Practices

8.2.1 Classification of Incentive Mechanisms

The incentive mechanisms classification presented in this section covers most known classes of incentives in general use in different types of human organizations – companies, non-profit (voluntary) organizations, engineering/design teams and crowdsourcing systems. Different organizations employ different (combinations of) incentive mechanisms to stimulate specific responses from agents. The classification is derived by the authors from: a) a multidisciplinary review of relevant domain literature cited throughout this section; and b) a survey of existing practices in social-computing platforms presented in Section 8.2.4.

- **Pay-per-performance (PPP)** – PPP is one of the most commonly used incentive mechanisms. The guiding principle states that every agent should be compensated proportionally to his contribution. Labor types where quantitative evaluation can be applied are particularly suited to this mechanism.

  A typical representative of the PPP incentive is the *wage*. As shown in Equation (8.1), the wage (w) usually consists of a fixed compensation amount (salary, w_0) and a variable amount (w_{inc}). The variable amount depends on measurable signals (s_i). Every signal is scaled by its weight coefficient (λ_i). Coefficient values depend not only on the actual importance that the principal attaches to a particular signal, but also on the accuracy of measurement that can be achieved.

$$w = w_0 + w_{inc}$$

$$w_{inc} = \beta \cdot \left[\lambda_1 s_1 + \lambda_2 s_2 + \cdots + \left(1 - \sum_{i=1}^{n-1} \lambda_i \right) s_n \right] \tag{8.1}$$

As the Informativeness Principle suggests, the more signals we include, the more accurate the evaluation we obtain. However, each signal value contains an intrinsic, normally-distributed measurement error. The incentive designer needs to take this into account. A lower value of the coefficient reflects the inaccuracy of measurement. In addition, each signal measurement has costs associated with it. In theory, the additional money needed for paying the rewards is provided from the additional profits obtained from the increased productivity. Therefore, designing an effective PPP incentive requires a proper trade-off to be found between the costs of measurement and the accuracy obtained. A signal value can also represent a mark based on a subjective performance evaluation by a supervisor.

In practice, this type of incentive strategy shows significant, verifiable productivity improvements of 25-40% when used for simple, repetitive production tasks, both in traditional companies ([97]), as well as with Human Intelligence Tasks (HITs) on Amazon's Mechanical Turk platform ([109]). Other studies, cited in [136], conclude that about 30-50% of the productivity gain is due to the filtering and attraction of better workers, thanks to the *selection effect* of this kind of incentive. This is an important finding, because it explains why even with relatively small amounts of incentives it is possible to achieve higher profits. In fact, increasing the amount of incentives for the same effort over time can lead to the *anchoring effect*, causing the agents to overestimate personal qualities. Keeping them reasonably low enables the selection effect, while not producing the anchoring effect.

Problems that characterize the application of PPP include typically: measurement inaccuracy, choice of signals and multitasking (see Section 8.2.3). This is in accordance with the newly observed results from [109], which confirm that the quality of work does not increase if the productivity is the only signal evaluated. Additionally applying some aggregate measures of performance can help alleviate these problems. Another problem that may arise due to an implemented PPP scheme is decreased solidarity among workers, potentially hampering the transfer of know-how and experience among workers. Again, the countermeasures are similar to the ones for multitasking, especially team-based strategies including apprentice relations, where the team gains are related to the professional progress of novices.

As already explained, the context in which a particular incentive strategy is implemented can determine its effectiveness. As demonstrated in [70, 74] PPP may not be an appropriate strategy to choose in cases where the agents are highly interested in the quality of the output. That includes domain experts from almost any area, who due to their expertise can usually earn enough money, so their primary motivation is not the monetary reward but the quality of the product or the reputational gain. PPP is also not suited for large, distributed, team-dependent tasks, where measuring individual contributions is inherently difficult. However, it is frequently used to complement other incentive schemes.

- **Quota systems & Discretionary bonuses** – These mechanisms are tightly related to the PPP mechanism. The conditions for applying them are the same as in the case of PPP. We observe and measure the same signals. What is different is that instead of rewarding agents proportionally to their productivity, the principal sets a number of performance-metrics thresholds. When an agent reaches a threshold he is given a one-off, predefined bonus. Quota systems evaluate at predefined moments whether a performance signal surpasses a threshold (e.g., yearly bonuses). On the other hand, discretionary bonuses are paid whenever an agent reaches a performance level for the first time (e.g., upon reaching a landmark number of customers).

  Being exposed to the issues of inaccurate measurements, these incentive strategies also suffer from the appearance of multitasking. Additionally, two other phenomena have been observed [136]:

  – The amount of effort always drops after an evaluation if the agent perceives the time until the next evaluation as long enough;
  – When the performance level is close to an award-winning quota motivation is significantly higher than the motivation of agents who have already exceeded the quota or feel they have no realistic chances of achieving it (on time).

  Therefore, the evaluation intervals and the quotas should be set in such a way that they can be reachable with a reasonable amount of additional effort, albeit not too easily. It is clear that these two parameters are highly context-dependent, and therefore can be determined only after observing historical records of employee behavior in a particular setup. Ideally, these parameters should be dynamically adjustable.

- **Deferred compensation** – This mechanism is similar to a quota system, in that an evaluation is made at predefined points in time. The subtle but important difference is that deferred compensation takes into account three points in time (t_0, t_1, t_2). At t_0 an agent is promised a reward after successfully passing a deferred evaluation at t_2. The evaluation takes into account the period of time $[t_1, t_2]$ and not just the current state at t_2. In case $t_1 = t_0$ the evaluation covers the entire interval.

  Deferred compensation is typically used for incentivizing agents working on complex, long-lasting tasks. The advantage is that it allows a more objective assessment of an agent's performance from a time distance. At the same time, the agent is given enough time $[t_0, t_1]$ to adapt to the new conditions, and then to prove the quality of his work over a period of time $[t_1, t_2]$. The disadvantage of this mechanism is that it is not always applicable, since agents are not always in a situation to wait long periods for a significant part of their compensation. A common example of this mechanism is the referral bonus. A referral bonus is a reward for employees for recommending or attracting new, suitable employees or partners to the organization.

- **Relative evaluation** – Although this mechanism can have many variations, the common underlying principle is that an entity is evaluated with respect to other entities within a specified group. The entity can be an agent (human), or an artifact (movie, document, product). Relative evaluation is used mainly for two reasons:

  - By restricting the evaluation to a closed group of entities (individuals), it removes the need to set explicit, absolute performance targets in conditions where such targets cannot easily be set, due to the dynamic and unpredictable nature of the environment.
  - It has been empirically proven that people respond positively to competition and comparison to others. (e.g., in [173]).

  Much of the initial success of Amazon and eBay can be attributed to the usage of good *reputation systems* [142], which in turn rely on relative evaluations of products by the customers.

- **Promotion** – Empirical studies [177] confirm that the prospect of a promotion increases motivation. A promotion is the result of competition for a limited number of predefined prizes. The prize is usually a higher position in the organization's hierarchy, bringing along higher pay, more decision-making power and more respect and esteem, although other prizes are also possible. Often, the benefits enjoyed by the agent after a promotion are disproportionately higher after a promotion compared to the benefits in the previous position. The reason for this is not to reward fairly the person currently holding the position, but rather to make future contenders for that position more competitive. In fact, the more an agent moves up the hierarchy, the more the rewards become disproportionate to personal abilities and productivity, moving away from PPP principles and focusing on competitiveness.

  Promotion is usually treated under the tournament theory ([98]), although other models also exist. The advantage of promotions is that they also eliminate centrality bias and force positive selection, as management cannot select inappropriate persons to advance, as that would mean transferring a great responsibility to unreliable persons, and ultimately produce greater costs to the principal. The drawback is that by valuing individual success, the method's application can de-motivate agents from helping each other and engaging in collaborations. Promotion often incorporates subjective evaluation methods, although other evaluation methods are possible.

Mechanism	Usage environments			Application considerations			
	Traditional Company		Social Computing	Positive application conditions	Negative application conditions	Advantages	Disadvantages
	SME	Large enterprise					
Pay-per-performance	++	+++	+++	quantitative evaluation possible	large, distributed, team-dependent tasks; measurement inaccuracy; when favoring quality over quantity	fairness; effort continuity	oversimplification: decreased solidarity among workers
Quota Sys. / Disc. Bonus	+	+++	+	recurrent evaluation intervals	when constant level of effort is needed	allows peaks/intervals of increased performance	effort drops after evaluation
Deferred Compensation	+	+++	+	complex, risky, long-lasting tasks	subjective evaluation; short consideration interval	better assessment of achievements; paying only after successful completion	workers need to accept taking the risk and waiting for compensation
Relative Evaluation	+	++	+++	cheap group evaluation method available	subjective evaluation	no absolute performance targets; eliminates subjectivity	decreases solidarity; can discourage beginners
Promotion	++	+++	+	need to elicit loyalty and sustained effort; when subjective evaluation is unavoidable	flat hierarchical structure	forces positive selection; eliminate centrality bias	decreases solidarity
Team-based Compensation	+	++	+	complex, cooperative tasks; inability to measure individual contributions	when retaining the best individuals is priority	increases cooperation and solidarity	disfavors best individuals
Psychological	+	+	++	stimulate competition; stimulate personal satisfaction	when cooperation needs to be favored	cheap implementation	limited effect on best and worst workers (anchoring effect)

Table 1: Left: Adoption of incentive mechanisms in different business environments (+ : low, ++ : medium, +++ : high). Right: Different application considerations

- **Team-based compensation** – This mechanism is used when the contribution of individual agents in a team environment cannot easily be identified. With this mechanism, the entire team is evaluated and rewarded. The reward is then split among the team members. Team-based compensation is susceptible to different dysfunctional behavioral responses. Worse-performing agents are effectively hiding within the group. At the same time, the performance of the better-performing agents is "diluted". Furthermore, teams often exhibit the *free-rider phenomenon* [136] – a situation in which individuals waste more resources (time, money, materials, equipment) than they would if individual expenses could be measured. The consequence is that the total expenses of a team surpass the summed expenses of independent individuals. Minimizing these negative effects is the primary challenge when applying this mechanism [85]. The most common variants are *team-level compensation* and *profit sharing*.

  When team-level compensation is used, the entire team is treated as an individual. After evaluation, the team is compensated by a (usually) monetary reward, which is then equally split among all team members. However, the scenario in which a reward is equally divided among members can lead to the dysfunctional behaviors we described above. In some cases, the better-performing team members will themselves naturally exert pressure on the free-riders, and thus weaken their negative effects. However, in cases where this does not happen, an attempt to differentiate individual efforts can be made. Peer voting is the most effective group evaluation mechanism in such cases, and it may be employed to differentiate agents and split the reward accordingly. This is clearly an example of a hybrid approach combining the idea of a team-based incentive, together with an incentive strategy targeted at individuals to eliminate dysfunctional behavior. Some studies (e.g., [134]) have shown that hybrid incentive strategies are indeed more effective than pure team-based compensation.

  The decision on the reward amount can be a matter of subjective or quantitative evaluation. Even with a constant high level of effort, the performance of the team can vary throughout its lifetime, depending on the compactness and interconnectedness of the group and the task that the team is working on. So, finding appropriate reward amounts becomes a difficult task [71]. One way to avoid having to decide on the amount of compensation in cases of unknown outcome of the collaborative effort is to tie it to the profit the company (or a company section) makes. This strategy is called *profit sharing*.

  Quantitative or subjective evaluation is usually used, often in combination with peer voting. The incentive action is usually a monetary reward, divided among team members equally or according to individual ratings.

- **Psychological incentive mechanisms** – Psychological incentives are the most elusive, making them hard to define and classify, since they often complement other mechanisms or even occur within them. They are mostly meant to target the intrinsic motivation of individuals. They can be operatively described as mechanisms that must: a) relate to human emotions; b) be advertised by the principal; c) be perceived by the agent.

The already mentioned incentive strategy of Stack Overflow, apart from being an example of a relative evaluation strategy at the same time employs a number of psychological incentive mechanisms, such as: status, points or badges. They serve not only to attest to the quality of contributors and answers, but also to motivate further contributions. If the points and statuses were not shown to the others and advertised, but rather used for evaluation only, we would still have indirect evaluation, but would miss out on the possibility to motivate users. Similarly, psychological incentive mechanisms can be coupled with a quota system. We already mentioned how an agent's productivity/motivation rises upon nearing a bonus quota. Even though an agent is well aware of the quota he is trying to achieve, the principal nonetheless advertises how "little" it is still left to achieve the goal to further boost the agent's motivation. Acting upon human fear is also a tactic commonly (mis)used (e.g., threat of dismissal or downgrading). The threat of being dismissed or downgraded is a powerful motivator, although very stressful for agents and causing different types of unforeseeable dysfunctional behavior.

Perception of the incentive by the agent affects its effectiveness. As the perception is context-dependent, choosing an adequate way of presenting the incentive is not a trivial decision. For example, choosing and advertising the employee of the month in societies where the sense of common good is highly valued can be very effective. In more individually oriented environments it is competition that drives performance. A principal may choose to exploit this fact by showing performance comparisons to (targeted) agents.

Psychological incentives have long been used in video games to elicit player dedication and motivation. Today, the same techniques (gamification) are used to make boring tasks (product reviews, customer feedback) appear more interesting and appealing. As a large number of business models of Internet-based companies depend on the revenues obtained through placing targeted advertisements, incentivizing customers to provide accurate product reviews and leave feedback becomes fundamentally important.

8.2.2 Composition of Incentive Mechanisms

In practice, employing a single incentive mechanism is usually not enough. Most organizations need to combine different incentive strategies to target different work roles and employees with different statuses. If we look at an engineering company, it is quite common for it to be organized in teams at the lowest level. Such a company would typically have teams of engineers developing the products, testing teams, marketing teams, sales teams, IT teams, customer support teams, etc. In addition, there are employees responsible for providing other services necessary for the running of the company (finance, HR, security, transport, supplies). A number of teams forms a unit responsible for a family of related products, or a number of related projects. Different teams are led by managers, who in turn respond to higher-positioned managers (e.g., project managers, product managers, division managers). Within each team engineers can have ranks, such as: junior engineer, senior engineer or distinguished engineer. More experienced engineers usually act as team leads, i.e., highly experienced professionals who drive the technical aspect of product development.

Such a company may decide to use a profit-sharing scheme at the division level, i.e., reward with all the teams in the division by a profit share if the financial results of the division are positive. In addition to that, individual efforts within a team can be stimulated by individual incentive strategies. In some teams where the contribution can be easily established (e.g., number of customer cases solved in case of customer support) PPP may be used to filter and keep the best employees. In other teams, we can use a combination of the subjective evaluation and peer voting to assess the contributions and adjust the variable part of the salary. Team members are given an opportunity to advance in rank and into managerial positions and keep advancing further if they accomplish certain goals. Every promotion brings along an increase in pay but also in responsibility. Top managers are evaluated exclusively with respect to the success of the company, and the payment of bonuses may be deferred. Additionally, the company may decide to give out referral bonuses, and award "employee of the year" awards.

Composing incentive mechanisms is often not simply wanted by an organization to improve performance, but also required to prevent dysfunctional behavior (possibly arising from the application of previous incentive mechanisms). At the end of Section 8.2.4.2 we describe the former incentive strategy of the company Locationary, which nicely showcases this: the originally introduced PPP was causing too many non-profitable contributions by crowd workers; subsequently a profit-sharing team-based mechanism was composed into the scheme to influence the quality of the contributions.

Table 1 presents a condensed view of different usage environments and application considerations of the incentive mechanisms we described.

8.2.3 Identifying Constituent Parts of Incentive Mechanisms

The related work we analyzed (Chapter 6.3) has not gone past the level of granularity of incentive mechanisms. We believe that this in great measure prevents development of generic handling of incentives in information systems. The goal of this section is to identify finer-grained building elements that can be individually modeled and used in information systems to compose and encode incentive mechanisms.

By analyzing the previously described incentive mechanism categories in Section 8.2.1 we can identify the following *incentive elements*, i.e., the atomic subcomponents in terms of which all mentioned incentive mechanisms can be expressed (Figure 8.3):

1. **Evaluation method** – provides inputs (signals) on agent performance to the incentive mechanism. Those inputs are evaluated in the logical context defined in the incentive condition.

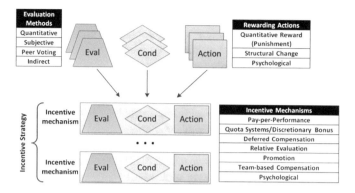

Fig. 8.3: An entire incentive strategy of an organization can be composed using smaller, modelable components – incentive elements

2. **Incentive condition** – contains the business logic of the incentive mechanism, i.e., the logical rules for application of certain rewarding actions.
3. **Rewarding action** – a concrete activity exerted upon targeted agents meant to influence their *future* behavior. Represents the outcome of the incentive mechanism.

8.2.3.1 Evaluation Methods

Individual Evaluation Methods

As the name suggests, these methods are used to evaluate agents individually, i.e., not explicitly conditioning their scores with the scores or opinions of other agents.

Quantitative evaluation represents the rating of individuals based on measurable properties of their contribution. Quantitative evaluation is attractive because it does not require human participation and can be entirely implemented in software. It is considered to be a precise and fair method. However, as it is not suitable for all purposes, it is often combined with other methods.

Some labor types are suited to precisely measuring the individual contributions of an agent (e.g., OCR correction, image labeling). In this case the agent can simply be evaluated on the number of units processed. But apart from the most primitive labor types, evaluation of an agent's performance requires evaluating different performance aspects (i.e., measurable signals), the most common being productivity, effort and quality of product. Different metrics are usually taken into consideration with different weights, depending on their importance and measurement accuracy. For example, in case of a product assembly line, the metric can be the number of units assembled, but also (with lower importance) the quality of assembled products, since the quality of work of a particular worker cannot always be precisely established. In other cases, e.g., in case of telemarketing where different phone agents are covering different neighborhoods, towns or ethnic groups, effort of agents may be highly valued compared to the number of units sold, because the success of sales of a particular product may depend on the geographical location of the area, wealth, climate or local habits. Effort is always a highly valued metric in cases where agents are not working under equal conditions.

Problems that arise here are measurement inaccuracy and the difficulty of choice of proper signals and weights. An additional problem is the phenomenon called *multitasking*. In spite of its counterintuitive name, it refers to agents putting most of the effort into tasks that are subject to incentives, while neglecting other tasks, subsequently damaging overall performance [76]. The principal can fight this kind of misbehavior by additionally employing subjective evaluation.

Subjective Evaluation Many aspects of human work are not quantifiable. The reasons can be:

- there are no clear outputs to evaluate;
- contribution has properties understandable and valuable to humans only;
- tasks are too complex to be clearly defined.

For example, whether a logo design is good or not is ultimately a matter of the aesthetic preference of the customer. In such cases we need to substitute an objective measurement with a human, subjective assessment of the quality of the work. In this case a human acts as a mapping function that quantifies human-oriented work aspects by wrapping together all the undefinable signals into one subjective assessment signal. Subjective evaluation is a widely used evaluation mechanism. Its advantages are simplicity and low cost, but its implementation as a human-based task makes it inherently imprecise and prone to dysfunctional behavioral responses.

Some of the phenomena that characterize this evaluation method that have been observed in practice [136] include:

- *Centrality bias* – ratings concentrated around some average value. Not enough differentiating of "good" and "bad" workers.
- *Leniency bias* – discomfort at rating "bad" workers with low marks.
- *Rent-seeking activities* – actions taken by employees with the particular goal of increasing the chances of getting a better rating from the manager, often including personal favors or unethical behavior.
- *Embellishment* – tendency of managers to rate subordinates better than deserved if the manager's own reputation or team-bonus depend on it.
- *Theft* – tendency of managers to consistently give lower scores to subordinates to save budgeted money if the compensation of the employees depends on the scores.

Centrality bias and leniency bias can be prevented by checking whether the ratings follow a distribution with specified parameters. However, this is not always preferable in practice, because it could motivate managers to perform data fitting. The usual solution for these, and also for other listed problems, is to make the ratings of managers subject of incentives as well. In practice, it means punishing a manager if his ratings of individual subordinate employees significantly or consistently differ from the ratings of other managers who worked with them, or from the ratings of their co-workers.

Group Evaluation Methods

Group evaluation methods evaluate an agent by aggregating assessments of community members.

Peer Evaluation (Peer Voting) is an expression of collective intelligence where members of a group evaluate the quality of other members. In the ideal case, the aggregated, subjective scores represent a fair, objective assessment.

The better the voters know the object of the vote, the better they can judge it. If the voting group is large enough, this method eliminates or alleviates the problems that subjective and quantitative evaluation suffer from. Centrality and leniency biases are alleviated by the fact that the votes will be better distributed, as the aggregated scores cannot be subjectively influenced. Since there is no longer a single voter who decides, activities that target a single voter's interests, such as embellishment, theft and rent-seeking are eliminated. Since a large number of different and professional peers evaluates different performance aspects, that leaves less space for multitasking activities.

This method also suffers from different weaknesses. In small, interconnected groups the voters can be unjust or lenient because of personal reasons. They can also feel uncomfortable and exhibit dysfunctional behavior if the person being judged knows their identity. Therefore, anonymity is often a favorable property in such cases. Another way of fighting these dysfunctional behaviors is to make voters subject to incentives: votes get compared, and those that stand out are discarded. At the same time one keeps track of agents' voting history to prevent consistent unfair voting.

When the community consists of a relatively small group of evaluated persons and a considerably larger group of voters, and both groups remain stable throughout the time, use of this method is particularly favorable. In that case, the voters have a good overview of much of the evaluated group. Since the relation voter-evaluated is unidirectional and will probably not change over time, voters do not have interest to exhibit dysfunctional behavior. This pattern is very common on the Internet today.

The method works as long as the size of the evaluated group remains small enough. As the evaluated group grows, voters become unable to keep up and acquire all the new facts necessary to pass fair judgments. Then they opt to rate better those persons or artifacts they know or feel traditionally have a good reputation ([137]). This phenomenon is known as *preferential attachment*, or colloquially "the rich get richer." It can be noticed on news sites that attract high numbers of user comments. Newly arriving readers usually tend to read and vote the most popular comments only, leaving many interesting comments practically unevaluated. Therefore, determining the group of voters and evaluated agents is crucial when designing instances of this incentive mechanism.

In traditional businesses, the major obstacle to applying this method was the cost, both in time and in money. Additionally, it was technically challenging, if not impossible, to apply this method often enough, and with appropriate voting groups. However, use of information systems, the Internet and social networks permitted a drastic decrease in application costs. A number of implementations already exist on the Internet (e.g., Like-button, binary voting, star voting, polls) but we lack a unified model able to express their different flavors and specify the voters and evaluated groups. Again, the act of voting is modeled as a human task, requiring active human participation.

Indirect Evaluation Since human performance is often difficult to define and measure, it is common to evaluate humans based on properties of and relations among the artifacts they produce. As the artifacts are always produced to be consumed by others, the decision on their quality is left to the community.

The artifacts are connected by various relations among themselves (contains, refers-to, subclass-of etc.), as well as with users (e.g., author, owner, consumer). The method of mapping properties and relations of artifacts to scores is non-trivial in the general case. An algorithm tracks relations and past interactions of the agent or his artifacts with the artifact that is being evaluated and calculates the score. For example, in [83, 152] the authors evaluate users of peer-to-peer networks by monitoring the content contributions of the users. Similarly, scientists can be evaluated by the number of their publications in journals, which in turn are ranked by their impact factor, which depends on the number of citations scientists make. The well-known e-labor (freelance) platform UpWork[3] uses a proprietary algorithm for worker evaluation and ranking [40]. Usually, a tailor-made algorithm needs to be developed, or an existing one adapted to a particular environment. The major difference from peer evaluation is that here the agent does not actively evaluate the artifact, and hence the algorithm is not dependent on interacting with the agent.

Another efficient method is by employing peer voting to evaluate artifacts. If we have favorable conditions for applying low-cost peer evaluation on artifacts then we eliminate the problem of dummy artifacts. It is also an added value for the accuracy of the algorithm. As the conditions for applying peer evaluations within Internet communities are usually favorable, this is a very commonly employed technique today.

Advantages and drawbacks of this method fully depend on the properties of the particular algorithm. If the algorithm is suitable it will exhibit fairness and prevent false results. The cost of this method also depends on the costs of developing, implementing and running the algorithm. A common problem is that users who know how the algorithm works may try to deceive it by outputting dummy artifacts with the sole purpose of increasing their scores. Detecting and preventing such attempts requires the algorithm to be amended, further increasing the costs.

Table 2 lists some common application and composability considerations for evaluation methods presented here. It also indicates how drawbacks of a particular evaluation method can be alleviated by combining it with other methods.

---

[3] https://www.upwork.com/ Result of merger between oDesk and Elance.

Evaluation Methods		Application considerations			Composability			
		Advantages	Disadvantages	Active Human Participation	Issues	Alleviated by	Solving	Typical Usage
Individual	Quantitative	fairness, simplicity, low cost	measurement inaccuracy	no	multitasking	peer evaluation; indirect evaluation; subjective evaluation	subjectivity-caused issues;	PPP; Quota Systems; Promotion; Deferred Compensation
Individual	Subjective	simplicity, low cost	subjectivity; inability to assess different contribution aspects	yes	centrality bias; leniency bias; deliberate low-scoring; embellishment; rent-seeking activities	incentivizing the decision maker to make honest decisions (e.g. by peer evaluation)	multitasking	Relative Evaluation; Promotion;
Group	Peer	fairness; low cost in social computing environment	active participation required	yes	preferential attachment; coordinated dysfunctional behavior of voters	incentivizing the peers (e.g. also by peer evaluation)	multitasking; subjectivity-caused issues	Relative Evaluation; Team-Based Compensation; Psychological
Group	Indirect	takes into account complex relations among agents and their artifacts	evaluation algorithm cost of development and maintenance	no	depending on the algorithm in use; fitting data to the algorithm	peer voting; better implementation of the algorithm	subjectivity-caused issues; peer evaluation issues	Relative Evaluation; Psychological; PPP

Table 2: Application and composability considerations for evaluation methods

8.2.3.2 Rewarding Actions

In order to induce a future specific behavioral response from agents the principal must perform one or more **rewarding actions** towards them.[4] The application of the actions is often colloquially called *rewarding* or *incentivizing*. A rewarding action can be one of the following types:

- Reward
- Structural change
- Psychological action

Rewards (but also: penalties, fines) can be modeled as quantitative changes in parameters associated with an agent or a group of agents. For example, a parameter can be the wage amount, which can be incremented by a bonus, or decreased by a penalty. Similarly, a parameter can be an agent's status, or a collection of objects in the agent's possession (e.g., FourSquare motivates users by assigning them different badges for different check-in patterns)

Structural changes are an empirically proven [98] motivator. A structural change does not imply strictly positional advancement/downgrading in traditional tree-like management structure. It also includes belonging to different teams at different times or collaborating with different people. For example, working in a team with a distinguished individual can diversify an agent's experience and boost his career. One way of modeling structural changes is by graph rewriting [14].

Psychological actions. Although all incentive actions work by exerting a psychological effect, what we denominate as psychological actions are only those in which an agent is influenced purely by being presented with some information. For example, we may decide to show an agent only the results of a couple of better-ranking agents rather than the full rankings. That way, the agent will not know his position in the rankings, which can be beneficial in two ways – by preventing the "anchoring effect" [109] for agents in the top part of the rankings and by preventing discouragement of agents in the lower part. Psychological actions do not include any explicit parameter or position change, but the diversity of presentation options means that defining a unified model for describing different psychological actions is still an open challenge. Effects of these actions are hard to measure precisely, but apart from empirical evidence [58], their broad adoption on the Internet today is another clear indication of their effectiveness.

Apart from the type of the rewarding action, another crucial aspect of the action's efficacy is the *timing* of the action (Figure 8.4). We can distinguish the moments: 1) when the action is announced/advertised to the agent; and 2) when the action is applied. The period between the two moments can be used to evaluate agent signals. The period spanning from the moment of the announcement and lasting possibly for an unspecified amount of time, but at least until the moment of the application of the action is called the *effectiveness range*.

---

[4] A punishment is simply a term used to describe a rewarding action meant to prevent a specific behavior instead of inducing one.

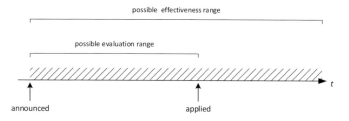

Fig. 8.4: Application and effectiveness of rewarding actions

8.2.3.3 Incentive Conditions

Incentive conditions state precisely how, when and where to apply rewarding actions. Each consists of at most three components (subconditions):

- *Parameter component* expresses a subcondition in the form of a logical formula over a specified number of parameters that describe an agent. For example, such a condition could filter out all the agents whose productivity is less than the team's average.
- *Time component* is used to formulate a condition over past behavior of an agent. For example, select all the agents who within the last three months had an unsatisfactory productivity level.
- *Structure component* filters out agents based on the relations they take part in. This component can be used to select members of a team, or all the collaborators of a specific agent.

By using all three components at the same time we can specify a complex condition, e.g., "incentivize the subordinates of a specific manager, who over the last year achieved a score higher than 60% in at least 10 months."

Incentive conditions are part of the business logic, and as such are stipulated by the domain experts empowered by the principal to manage the workforce. However, a small organization can take advantage of some good practices and employ pre-made incentive models (patterns) adapted to fit the particular organization's needs. Feedback information obtained through monitoring execution of rewarding actions can be used to adapt condition parameters.

8.2.4 Incentive Mechanisms in Real-World Social Computing Platforms

8.2.4.1 Survey Criteria

In 2012[5] we surveyed over 1600 Internet-based companies and organizations that describe themselves using keywords such as "social computing" or "crowdsourcing". We investigated their business models and contracts offered to users/participants/workers, as described on organizations' websites. The main goals of the survey were:

- To demonstrate that the classification we had established mostly based on our multidisciplinary literature survey is valid and applicable also for internet-based social-computing business models.
- To gain a better insight into the usage patterns of different incentive mechanisms, evaluation methods, rewarding actions and combinations thereof.

We filtered the companies in such a way to exclude those that fulfill any of the following criteria:

- Crowd-funding websites, humanitarian & community-benefit sites or voluntary-contribution sites.
- Sites that only act as intermediaries to establish links between human service providers and consumers (often charging commission for providing secure transaction environment), except when they employed incentive mechanisms to increase the number and quality of participants.
- Sites that just provide a technical solution or environment to do the business.
- Sites that directly sell products created/owned by the crowd – e.g., stock photography.
- Sites that do not disclose or clearly/publicly state the incentive scheme.
- Sites that were not in one of the following languages: English, German, Spanish, Portuguese, Italian.

We also decided not to include companies employing gamification approaches in the classification part of our survey. There are several reasons for this decision:

- Every gamification approach can be considered a psychological strategy, with quantitative evaluation.
- A company rarely bases its principal incentive strategy on gamification only.
- Many companies employing gamification are not primarily social-computing companies, but traditional companies relying on gamification elements to attract users to perform uninteresting tasks.

---

[5] Please note that some information in this section may be outdated due to the highly dynamic nature of the social-computing market.

However, we acknowledge the growing importance of gamification approaches. For example, SAP (within their SAP Community Network[6]) enables their employees to build up reputation by writing articles, guides and samples, answering questions etc. That not only helps other team members get useful information more easily, but also raises the reputation of the employee that transfers the knowledge. The contributors are scored, and score boards are publicly available. Even though the score does not bring any concrete rewards, the reputation gained can implicitly bring better career advancement opportunities and higher respect from colleagues.

Another example of how gamification can incentivize employees is the use of games in teaching employees to better understand, use and represent the products of their own company (e.g., IBM Innov8[7]). While employees get distracted from dull tasks for some time and learn something in the process, the company gets better-skilled and more competitive workers. Finally, yet another gamification example can be seen in a project run by the National Library of Finland[8], where contributors are engaged in a game with the true purpose of correctly recognizing scanned material from the library archives.

8.2.4.2 Survey Results

After applying the filtering rules stated above, out of over 1,600 examined companies we identified 140 companies (8.75%) that employed and clearly described the rewarding/incentive practices (types of awards, evaluation methods, rules, conditions). We then classified them according to the previously described classification (Section 8.2.1).

The most surprising finding was that 59 of the 140 companies (42%) employed a very simple "contest" business model employing a relative evaluation incentive mechanism, meaning that a creative task is deployed to the crowd. Each crowd member (or entity) then submits a design. In the vast majority of cases the best design is chosen by subjective evaluation (85%). That was expected, since the company buying the design reserves the right to ultimately decide on the best design. In fact, in many cases, it is the only possible choice. The remaining "contest" companies employ peer evaluation (10%) or quantitative evaluation (5%). When using peer evaluation, the company delegates the decision on the best design to the crowd of peers, while taking the risk of producing and selling the design. In some cases, e.g., programming contests, the artifacts are evaluated quantitatively, by automated testing procedures. It is worth noticing that using peer or quantitative evaluation produces quantifiable ratings of the users. In such cases, individuals are better motivated to take part in future contests even if they feel they cannot win, because they can use their ranking as a personal quality proof when applying for other jobs or just as a matter of prestige. We expect to see an increase in the latter two evaluation categories,

---

[6] http://scn.sap.com/

[7] https://www.ibm.com/software/solutions/soa/innov8/

[8] http://www.digitalkoot.fi

as they help improve the quality of designs if the crowd is large, contains quality individuals, and is properly motivated. However, building up and managing such a crowd also implies the use of other incentive mechanisms. The contest model alone dissuades good (but not the best) agents, who rarely win the contests.

Apart from the 59 organizations running contests, relative evaluation is used by another 16 organizations, usually combined with various other mechanisms. This makes relative evaluation by far the most widely used incentive mechanism on the social-computing market today (54%) (Table 3). This is in contrast with its use in traditional businesses, where it is used considerably less [9], as the implementation costs are much higher.

Incentive Mech. Type	No. of Companies	Percentage
Relative Evaluation	75	54%
Pay-per-Performance	46	33%
Psychological	23	16%
Quota Sys. / Disc. Bonus	12	9%
Deferred Compensation	10	7%
Promotion	9	6%
Team-based Compensation	3	2%

Table 3: Use of incentive mechanism categories by social-computing companies

The other significant group are the companies that pay agents for completing human microtasks. We found 46 such companies (33%). Some of them are general platforms for submitting and managing any kind of human-doable tasks (such as the emblematic Amazon Mechanical Turk[9]). Others offer specialized human services, most commonly: writing reviews, locating software bugs, translating or performing some simple, location-based tasks, etc. What all those companies have in common is the use of a pay-per-performance mechanism (PPP). The tasks range from very simple (in majority of cases) to more imaginative and complex, such as locating bugs. Quantitative evaluation is the method of choice in most cases (65%). Quantitative evaluation sometimes produces a binary output, e.g., when submitting successful/un-successful steps to reproduce a bug. The binary output allows only two levels of the quality of work to be expressed, so the agents are rewarded on a per-task basis for every successful completion. In that case, the company usually requires no entry tests for joining the contributing crowd. In other cases, the quality of work is not easy to establish and the output is proportional to the quantity of finer-grained units performed (e.g., word count in translation tasks) but the agents are usually asked to complete entry tests. Pay rate for subsequent work is determined by the test results. Other evaluation methods include subjective and peer/indirect evaluation, both at 17%. It is interesting to note that peer evaluation for double checking results is not frequently employed, as companies find it cheaper to test the contributors once and trust their skills later on. However, as companies start to offer more complex human

---

[9] http://www.mturk.com/

tasks, quality assurance becomes imperative, so we expect so see a rise in peer and indirect evaluation. Eleven companies combine pure PPP with other mechanisms.

Only three companies employ a combination of four or five different mechanisms (Table 4). The most famous of them is uTest.com. As their business model requires them to have a large crowd of dedicated professionals, it becomes clear why they employ more than just simple PPP.

ScalableWorkforce.com is the only company in our study that advertises the importance of *crowd (workforce) management*. They offer the tools for crowd management on Amazon Mechanical Turk to their clients. Their tools allow for tighter agent collaboration (creating a sense of community among workers), workflow management, performance management and elementary career building.

No. of Inc. Mech.	No. of Companies	Percentage
1	116	83%
2	15	11%
3	6	4%
4	3	2%

Table 4: Number of incentive mechanisms used by social-computing companies. Over 80% of the companies employ only one mechanism

Twelve companies (8.5%) rely uniquely on psychological mechanisms to assemble and improve the agent community. The common trait is reliance on the indirect influence of rankings in the agent's (non-virtual) professional life. For example, avvo.com attracts large communities of doctors and lawyers in the US who offer free responses and advice to people visiting the website. Quality and timeliness of professionals' responses affect their reputation rankings, which can be used as a prestige advertisement to attract actual paying customers to their private practices. Another very interesting example is companies such as crowdpark.de or prediculous.com. They ask their users to "predict" the future by placing bets on upcoming events using virtual currency. Users that have the best predictions over time earn virtual trophies (badges), which is the only incentive for people to participate. The crowdsourced odds can be used to adjust odds in real betting.

Team-based compensation was used by only three companies we surveyed. For example, mercmob.com encourages formation of virtual human teams for various tasks. An agent expresses confidence in the successful completion of a task by investing part of a limited number of his "contracts". Once invested, the contracts are tied to the task, motivating the agents that accept the task to give their best to self-organize in a team and attract others to accomplish the task. If in the end the task is completed successfully each agent gets a monetary reward proportional to the number of invested contracts.

Discretionary bonuses or quota systems are used by eleven companies (8%). However, they are always used in combination with another mechanism – most commonly PPP (64%), as is also the case in traditional companies.

Deferred compensation is used by 7% of the companies, and usually as the only mechanism employed. Bluepatent.com is a company that crowdsources the task of locating prior art for potential patent submissions. The agents (researchers) are asked to find and submit relevant documents proving the existence of prior art. Deciding on the validity and usefulness of such documents is an intricate task, and hence the decision on the compensation is delayed until an expert committee decides on it. Advisemejobs.com pays out classical referral bonuses to the agents who suggest appropriate job candidates.

Only 7% of the companies offer some kind of career advancements, combined with other mechanisms. As the crowd structure is usually plain, career advances usually mean a higher status, implying a higher wage. We have encountered only two cases where advancement also meant some kind of structural change, with an agent taking responsibility for leading or supervising lower-ranked agents (e.g., uTest.com). In traditional companies the decision on a promotion of an employee is usually a matter of subjective evaluation by his superiors. With promotion being the most commonly employed traditional incentive, subjective evaluation is then also the most commonly used evaluation method. However, if we take out of the picture the companies running creative contests, where the artistic nature of the artifacts forces the use of subjective evaluation, we see that in the world of Social Computing this trend has reversed. Subjective evaluation trails behind quantitative and peer evaluation (Table 5). This is explained by the fact that the use of information systems enables cheaper measurements of different inputs and setting up of peer-voting mechanisms.

Evaluation Method	No. of Companies	Percentage
Quantitative Evaluation	51	63%
Peer Voting + Indirect	35	43%
Subjective Evaluation	14	17%

Table 5: Use of evaluation mechanisms (excluding companies running creative contests)

A small number of companies employ a combination of different incentive mechanisms. Locationary[10] was a company that used agents spread around the world to expand and maintain a global business directory by adding local business information.

Their strategy combined a number of incentive mechanisms: 1) "lottery tickets" (a Quota system, also known as "conditional PPP"); 2) team-based compensation (based on the "shares" of added companies); 3) deferred compensation, based on the trust scores of the agents.

---

[10] The incentive strategy was acquired before the company was taken over and integrated by Apple, Inc. http://allthingsd.com/20130719/apple-acquires-local-data-outfit-locationary/. The original URL was http://www.locationary.com/

With every new entry added/corrected an agent wins "lottery tickets' that increase the chances of winning a reward in a lottery. However, there is a minimum quota of tickets that represents the condition to enter the draw (hence "conditional PPP"). Tickets are not tied to any particular directory entry. Agents are given different ticket amounts for different actions (adding, editing or verifying different directory entry fields). The number of tickets issued to an agent for editing an entry depends on how valuable the (accuracy of the) entry is to the company. For example, a Google street view URL is more valuable than the URL of the web page of the place. Similarly, fixing outdated/incorrect data is highly appreciated.

This mechanism incentivizes the increased activity of the agents, but also motivates them to cheat, as some people will start inputting invalid entries to increase their chances of winning. Deferred compensation is used to counteract this caused dysfunctional behavior. The agents are only allowed to enter the prize draws if (apart from the ticket quota) their trust score is high enough. The trust metric plays a crucial role here. Trust is proportional to the percentage of approved entries, and this metric discourages agents from cheating. Entries can be approved or disapproved only by other highly trusted agents (an example of peer evaluation). Trusted agents are motivated to perform validation tasks by getting more lottery tickets than they would get for adding/editing fields. On the other hand, cheaters are further punished by subtraction of lottery tickets for every incorrect data field they provided.

The incentive strategy described so far does a good job of attracting a high number of entries and keeping them fresh and accurate. However, it does not discriminate between the directory entries themselves. That means that it motivates agents as much to enter information on an insignificant local grocery store, as to enter/update information on a high-profile company. As Locationary used to rely on advertising revenues, that meant that an additional incentive mechanism attracting higher numbers of profitable entries needed to be included on top of the strategy described so far. The team-based compensation played this role. Locationary used to share 50% of the revenues originating from a directory entry with the agents holding "shares" of that entry. Shares were given to the people who were first to provide new/additional information on the entry. Again, cashing out was permitted only to the trusted agents.

This example demonstrates how different mechanisms are used to target different necessities, and how they need to be composed to achieve their full effect. In Table 6 we list some examples of companies employing different evaluation methods within different incentive mechanisms.

8.2.4.3 Survey Conclusions

With creativity contests and microtask platforms dominating the landscape of Social Computing today we see that the organizational structure of agents is usually flat or very simple. Hierarchies and teams of agents usually do not exist. In such an environment, most Social Computing companies need to use only one or two simple incentive mechanisms. Promotion, commonly used in traditional companies, is rarely found within Social Computing companies. The reason is the short-lived nature

	Quantitative	Subjective	Peer	Indirect
Pay-per-performance	mturk.com	content.de	crowdflower.com	translationcloud.net
Quota/Discretionary Bonus	gild.com		carnetdemode.fr	
Deferred Compensation	advisemejobs.com	bluepatent.com	crowdcast.com	
Relative Evaluation	netflixprize.com	designcrowd.com	threadless.com	topcoder.com
Promotion	utest.com	scalableworkforce.com	kibin.com	
Psychological Incentives	crowdpark.de	battleofconcepts.nl	avvo.com	
Team-based Compensation		mercmob.com	geniuscrowds.com	

Table 6: Examples of companies employing different evaluation methods (columns) within different incentive mechanisms (rows) at the time the survey was compiled. Note: mechanisms presented here may not represent the only or primary mechanisms that the company uses

of transactions between agents and the Internet companies. For the same reason, team-based compensation is also poorly represented. The idea of building a "career in the cloud" is still considered in the theoretical domain.

On the other hand, most traditional companies combine different, elaborate mechanisms to elicit particular responses from agents and retain quality workers [136]. The mechanisms complement each other to mutually cancel out individual drawbacks. In many cases the (more complex) mechanism combinations arose only after practical use of simpler combinations exposed weaknesses which the agents would exploit to their benefit. We encountered such experiences also in the surveyed social-computing companies, with, e.g., the Locationary incentive scheme being a very illustrative example thereof.

Our survey shows that as the price of application of quantitative, peer and indirect evaluation has lowered, relative evaluation and PPP have become the most popular incentive mechanisms among social-computing companies. Subjective evaluation, although in total numbers well represented, is found largely within companies that base their business model on organizing creativity contests. Psychological incentives and gamification approaches are gaining ground. We expect them to achieve their full potential as amplifiers for other incentive mechanisms.

The envisioned growth in complexity of business processes and organizational structures for social computing will require novel, automated ways of handling the behavior of agent crowds. That is why we perceive a necessity to develop models of incentive mechanisms and incentive management frameworks fitting existing business models and real-world socio-technical systems, capable of supporting complex, composable incentive mechanisms.

Such frameworks need to be able to monitor worker crowds and perform runtime applications and adaptations of incentive mechanisms to prevent the diverse negative effects we described (e.g., free-rider problem, multitasking, biasing, anchoring, preferential attachment), switching when needed between different evaluation methods, rewarding actions and incentive conditions at runtime, while minimizing overall costs. This way, particular worker groups and behaviors can be efficiently targeted.

Additional benefits would include the following:

- Historical data can be used to detect performance bottlenecks, preferable team compositions, optimal wages etc. Additionally, we can make predictions and choose the optimal composition of incentive mechanisms for the future. This opens up the possibility of novel ways to achieve indirect, automated team adaptability through application of incentives.
- For certain business models, application of proven incentive patterns cuts costs in both time and money. The incentive patterns can be tweaked to fit particular needs based on feedback obtained via monitoring.
- By generalizing and formally modeling incentive mechanisms, we can encode them in a system-independent manner. That way, they become portable and reusable on different underlying systems, without new system-specific programming code having to be written.
- The management of rewarding and incentives can be offered remotely as a Web Service.

8.3 Modeling Incentives for Use in Socio-Technical Systems

This section builds on the first two sections to develop actionable incentive models that allow us to:

1. Model the previously described incentive mechanisms.
2. Model the application of incentive mechanisms from (a) and responses to it.

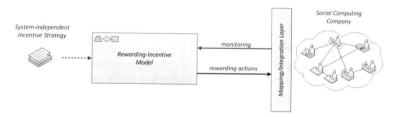

Fig. 8.5: A conceptual illustration of a system capable of translating portable incentive strategies into concrete rewarding actions for different socio-technical platforms

The models introduced in this section will allow us to build frameworks operating on these models, and offering the functionality of incentive management to third parties (Figure 8.5).

8.3.1 Comprehensive Incentive Model

Application of incentives always requires two interested parties - an *authority* (principal) and a *worker* (agent). The authority is interested in stimulating, promoting or discouraging certain behavioral responses in workers. The incentive exhibits its psychological effect by promising the worker a reward or a punishment based on the actions the worker will perform. The wish to get the reward or escape the punishment drives the worker's decisions on future actions. The reward (punishment) can be material or psychological (e.g., a change of status in a community – ranking, promotion). The type, timings and amounts of reward need to be carefully considered to achieve the wanted effect of influencing a specific behavior in a planned direction. In addition, introduction of incentives introduces additional costs for the authority, who hopes to compensate for them through the newly arisen worker behavior (e.g., increased productivity).

However, as soon as an incentive mechanism is introduced, it produces dysfunctional behavior responses in the worker population. The workers, being rational agents, adapt to the new rules and change their working patterns, trying to exploit the new incentive to profit more than the rest of the population. The authority compensates for this by introducing other incentive mechanisms targeting the dysfunctional behavior, further increasing the authority-side costs, and causing new types of dysfunctional behavior. However, once the proper combination of incentive mechanisms is put in place and calibrated, the system enters a stable state. The problem with crowdsourcing/social-computing processes is that the system may not stay long in a stable state due to an unforeseen change in worker participation or collaboration patterns. Therefore, the incentive setup needs to be reconfigured and re-calibrated as quickly as possible, in order to avoid incurring high costs to the authority. This feedback control-loop involving the authority and the worker represents the actual incentive mechanism that we are interested in modeling.

Modeling an incentive mechanism, therefore, always involves modeling both the authority and the worker side, as well as the possible interactions between them. In Figure 8.6 we show an abstract representation of the model of an incentive mechanism.

Workers differ from each other in having different sets of personal characteristics (e.g., accuracy, speed, experience). The characteristics are determined by a private set of variables stored in the *internal state S*. The internal state also contains records of worker's past actions. The internal state is private to the worker, and is used as one of the inputs for the *decision-making function f_a* that describes the worker's choice of the next action to perform.

In the majority of cases the internal state variables are normally distributed across the worker population. Occasionally, certain variables can be intentionally given predefined values to simulate a certain type of behavior, or a specific class of workers. This can also be used to simulate changes in behavior after unconstrained interactions among workers; for example, after an interaction with another worker a worker may be "persuaded" to decrease his performance levels by lowering his internal *effort*

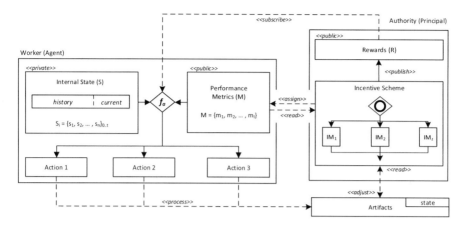

Fig. 8.6: Incentive mechanisms need to capture the interaction between workers (agent) and authority (principal)

metric. The algorithms modifying the internal state S are not prescribed by our model, and are freely chosen by the system designer.

Apart from the internal state, each worker is characterized by the publicly exposed set of *performance metrics M*, which are defined and constantly updated by the authority for each worker. The performance metrics reflect the authority's perception of the worker's past interactions with the system (e.g., trust, rank, expertise, responsiveness). Knowing this allows the worker to make better decisions on his future actions. For example, knowing that a poor reputation will disqualify him from getting a reward in future may drive the worker to work better or to quit the system altogether. It also allows him to compare himself with other workers. Therefore, the set of performance metrics is another input for the decision-making function f_a.

The third input for the decision-making function f_a is the set of promised *rewards (punishments) R*. Rewards are expressed as publicly advertised amounts/increments in certain parameters that serve as the recognized means of payment/prestige within the system (e.g., money, points, stakes/shares, badges). They are specified per action, per artifact, and per performance metrics (or a combination thereof), thus making them also dependent on a particular worker. For example, a reward may promise an increase of at least 100 points to the first/any worker who performs the action of rating an artifact. The number of points can then be further increased or decreased depending on the worker's reputation.

Workers interact with the authority solely by performing actions over *artifacts (K)* offered to the worker population by the authority. A worker's behavior can thus be described as a sequence of actions in time, $A_t \in A = \{A_0, ..., A_n\}$, interleaved with periods of idling (idling being a special case of action). The set of possible actions is the same for every worker. However, the effects of the execution of an action may be different, depending on the worker's personal characteristics from the internal state

S. For example, a worker with innate precision and more experience can improve an artifact better than a worker not possessing those qualities.

As previously stated, a worker's next action is selected through the use of a decision-making function $f_a = f(S, M, R)$ potentially considering all of the following factors: a) the statistically or intentionally determined personality of the worker; b) the historical record of past actions; c) the authority's view of the worker's performance; d) the performance of other workers; and e) promised rewards, with respect to the current state of one's performance metrics. The decision-making function is context-dependent, and defined by the model designer based on the observed/expected worker behavior.

From a social computing perspective, the authority's motivation for offering artifacts for processing to the worker crowd is to exploit the crowd's numerosity to either achieve higher quality of the artifacts (e.g., in terms of accuracy, relevance, creativity), or lower the cost (e.g., in terms of time or money). This motivation guides the authority's choice of incentive mechanisms. The authority has at its disposal a number of *incentive mechanisms* IM_i. Each one of them should be designed to target/modify only a small number of very specific parameters. Thus, it is the proper addition or composition of incentive mechanisms that allows the overall effect of an incentive scheme, as well as fine-tuning and runtime modifications.

An incentive mechanism *IM* takes as inputs: 1) the current state of an artifact K_i; 2) the current performance metrics of a worker M_j; and optionally 3) the output from another incentive mechanism returning the same type of reward, R'_{a_k}. The output of an incentive mechanism is the amount/increment of the reward R_{a_k} to offer to the worker M_j for the action a_k over artifact K_i.

$$IM : (K_i, M_j, R'_{a_k}) \rightarrow R_{a_k} \qquad (8.2)$$

The true power of incentive mechanisms lies in the possibility of their combination. The reward (f_R) can be calculated through a number of additions ($+$) and/or functional compositions ($\circ$) of different incentive mechanisms. For example, a worker may be given an increment in points for each time he worked on an artifact in the past. Each of those increments can then be modified, depending on how many other workers worked on that same artifact. In addition, the total increment in points can be further modified according to the worker's current reputation. The finally calculated increment value represents the promised reward. The set of finally calculated rewards per worker $R_w = \{f_{R_1}, ..., f_{R_z}\}$ is then advertised to the workers, influencing their future behavior, and closing the feedback loop. The major difficulty in designing a successful incentive scheme lies in properly choosing the set of *incentive parameters* (performance metrics, incentive mechanisms, and their compositions). Often, the possible effects when using one set of parameters are unclear at design time, and an experimental or a simulation evaluation is needed to determine them.

8.3.1.1 Definitions

Now we can define the key terms related to incentive management:

Worker (Agent)
  A human who is the target of incentives.
Authority (Principal)
  The entity engaging the workers for productive working purposes, administering incentives upon them.
Incentive
  Any activity or scheme employed by the authority to stimulate (motivate) an increased level of certain work-related qualities (e.g., productivity, speed, quality of work, number of participants) or to discourage certain activities (e.g., dropouts rate), before the actual execution of those activities.
Reward
  Any kind of recompense for worthy services rendered or retribution for wrongdoing exerted upon workers during the execution of the activity or after its completion. A reward can be made equivalent to an economic value (money or physical goods), or a social status such as prestige, rank, or expertise.
Incentive Mechanism
  A concrete rule for assigning/applying the rewards targeting a specific (group of) workers, based on certain logical, temporal and spatial criteria; A concrete implementation of an incentive for a given application context.
Incentive Element
  An atomic component (construct) in terms of which incentive mechanisms can be expressed.
Incentive Scheme
  The combined global effect of the application of a set of incentive mechanisms.

8.3.2 Rewarding Model (The PRINC Framework)

In the comprehensive model presented in the previous section, the authority reads a worker's performance metrics and the changes in artifact states associated with a worker as inputs for the incentive scheme. However, the model presented there says nothing further about how the authority is able to interpret those inputs and output concrete rewards. In this section, we investigate the authority's internal model for representing the workers, encoding incentive mechanisms and representing rewarding actions. The model is named the *Rewarding Model* (RMod), and presents the core component of the PRINC framework presented in [156].

For the authority (principal) the RMod represents the following aspects of a real-world incentive loop:

State
  Represents the quantitative state of the both the incentivized socio-technical

system as well as the internal business logic state needed for making incentive decisions. This includes global attributes and individual worker attributes representing different worker performance metrics (QoS).

Time

Expressed as a collection of time-annotated records of past and future worker interactions supporting various time conditions and constraints. The notion of timing is fundamental when dealing with incentives, as worker evaluation in most cases depends on the history of their past behavior. Similarly, a reward may be scheduled for a future moment if the performance metrics in the upcoming period meet the expectations.

Structure

Allows representation and manipulation of various types of relationships among workers. Workers are often stimulated just by being placed in a position to collaborate with people they find most comfortable to work with. In fact, proper team composition can be vital to process success, and can often be subject to changes during process execution. Finally, promotion, as one of the most widely used incentive mechanisms, implies a clear hierarchical change.

The authority employs a group of workers to perform a complex process, consisting of multiple tasks. It is assumed that the complete task lifecycle management (e.g., splitting into subtasks, task descriptions, task assignment, task negotiation and agreement) is under the control of the authority. A worker is assigned a (sub)task to perform in a given time and agrees to be the subject of incentive evaluations. Concretely, the authority and the worker agree that the worker may be subject to rewards/penalties in some predetermined cases. Workers can work individually on assigned tasks, in a formalized organization (team, collective) or relationship with the authority (e.g., be employed, be part of teams, have managers). The authority's entire knowledge about the progress of the task is obtained from periodic *messages* (updates) that it receives from the workers and subsequent reasoning about that data. The application of rewards to the workers is similarly abstracted as legally binding messages to worker.

A *task* is the basic working unit. Workers are rewarded for working on a particular task within the task's timeframe, although the outcome of the evaluation can also depend on the history of previous contributions. Therefore, the lifetime of a worker is not related to the duration of the task. The authority maintains its own view of the workers and the relations between them in a *community graph*. The nodes in the graph represent the workers, while the edges represent different real-world relationships among the workers. For example, they can represent records of past collaborations, notion of trust [165], dependencies, managerial relations, etc. In addition, each node is described by a set of attributes. The attributes may represent task-specific (short-lived) or permanent records of a worker's performance. This is the most general representation possible. However, in practice the model must be coupled with a real-world socio-technical platform, so the nodes and relations need to be mapped to entities in that platform.

The model assumes an iterative task execution. *Iteration* length is measured in clock ticks. A *clock tick* is the basic unit of time measurement. A worker's progress is read upon iteration expiry so the model can obtain up-to-date QoS metrics. One iteration is the basic time unit when monitoring and evaluating task execution. Iteration cycle length is tunable to allow better runtime adaptability, as the iteration length can be a significant factor when evaluating results and can affect the performance of the team.

In order to model history of past behavior, as well as scheduling of future performance evaluations and rewarding actions, we include in the model the notions of *timeline* and *event*. The timeline is a time-stamped collection of past and future event records. An *event* is an object encapsulating an executable action and a timestamp. Events are interpreted by the socio-technical platform as instructions or suggestions to the platform itself or particular workers. For example, an event could notify a worker that he needs to increase the QoS level of his service in future iterations, or face penalties. Similarly, it could instruct the platform to dissolve a team, invite new workers, or terminate contracts with others. Events can be generated by the platform itself or originate from an incentive mechanism in the RMod. They can target individual workers or groups of workers, depending on the query that forms part of the action contained in the event object. An event can also target global properties of the system itself.

An event can be in two states: scheduled and past. *Scheduled events* are used to enforce/influence future worker behavior. They contain information to execute performance measurements, evaluations or concrete rewarding actions at a specified moment in the future. Scheduled events can be canceled or re-scheduled when needed. The timestamp can be expressed either in iterations or in clock ticks. Time expressed in clock ticks is fixed, whereas time expressed in iterations is automatically recalculated to an appropriate clock tick if the iteration duration is altered. This can be useful in many real-world situations. For example, a Christmas bonus is to be paid out on a fixed date, while if a project stage is prolonged due to some unexpected events, we want to reschedule the current iteration and perform the rewarding only at its end. When the time to execute an event is reached, the contained action is executed and the results stored back in the event, which is then archived and put into past state. After that point, the purpose of the *past event* is to serve as a historical reference for future evaluations of workers. An event execution can generate new events, or perform modifications of the team structure and worker attributes. Events are initially generated by executing *rewarding actions*, which are part of the enforced incentive mechanisms. Figure 8.7 describes a typical working cycle of our RMod. Incentive mechanisms (IMs) provide the necessary logic for performing worker evaluations and rewarding actions. At every clock tick IMs get evaluated. Only the IMs that fulfill a logical condition will be triggered to execute. The IM examines the current state of the model, and if a rewarding action needs to be performed produces one or more event objects. The rewarding action contained in the event will include the business (incentive) logic specified in the IM. The events then get stored in the timeline. When the appropriate time comes, the events get executed, modifying the attributes and the graph structure, and possibly spawning new events. The *RModManager* boxes in Figure 8.7 represent the system that implements the functionalities and manipulates the RMod.

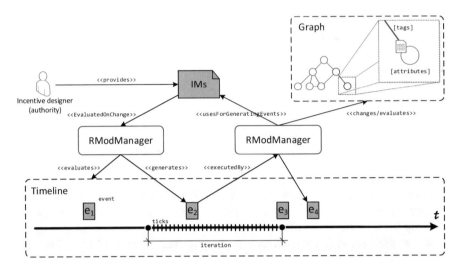

Fig. 8.7: Components and interactions in RMod

The RMod allows us to express various basic incentive mechanisms, such as:

- *"At the end of an iteration, reward each contributor who scored better than the average score of his neighbors in that iteration."*
- *"Reward every worker (contributor) who within the last* n *iterations scored a score* t *or greater in at least* k *iterations"* (k ≤ n).
- *"Assign the person with most check-ins at a place a 'Mayor' badge."*
- *"Unless the productivity increases to a level* p *within the next* n *iterations, replace the team's current manager with the most trusted of his subordinate workers."*

Furthermore, the model allows for easy composition of different incentive mechanisms, a feature necessary to target different dysfunctional behaviors of workers.

8.3.3 Evaluating RMod Through Simulation

The Rewarding Model (RMod) presented in the previous section represented the authority's simplified view of the workers in a socio-technical platform. The RMod is designed to be generic and simple enough to be able to encode most incentive conditions and rewarding actions described in Section 8.2.4. However, referring back to Figure 8.6 in Section 8.3.1, we see that in order to close the incentive loop we need also to be able to model and simulate the behavioral responses of workers to the applied incentives. Those responses are the result of the complex traits of human nature, the scenario-specific working environment and the social characteristics of the worker community. Therefore, in order to model and simulate behavioral responses, a context-specific model emphasizing selected behavioral traits needs to be developed each time anew. In this section, we present a *simulation methodology* for developing scenario-specific *simulation models* based on *agent-based social simulation*.

As shown in [51], most incentive mechanisms are developed based on empirical data obtained from different studies. However, the empirical findings are often context-specific, and when applied in different environments may yield different behavioral responses. This is especially true for incentive models that need to consider social characteristics of the worker community, such as coordinated group actions (e.g., worker resistance [115], informal forum-based worker coordination [107], social/regional/ethnic peculiarities, voluntary work [70], importance of reputation/flaunting [162], or web-scale malicious behavior [185]. An additional complication is that these phenomena change often and characterize different subsets of the crowd differently in different moments. This makes development of appropriate mathematical incentive models difficult.

The major problem an incentive designer is faced with in this case is how to evaluate the developed incentive mechanisms aimed at targeting disruptive or dysfunctional behaviors. The designer needs to consider factors, such as emerging, unexpected and malicious worker behavior, incentive applicability, range of stability, reward fairness, expected costs, reward values and timing. Failing to do so leads to exploding costs and work overload, as the system cannot scale with the extent of user participation typical of social-computing environments. Unbalanced rewards keep new members from joining or cause established members to feel unappreciated and leave. Ill-conceived incentives allow users to game the system, prove ineffective against vandalism, or assign too many privileges to particular members, tempting them to abuse their power.

In this section we present a methodology for incentive designers for quickly selecting, composing and customizing existing, real-world atomic incentive mechanisms, and roughly predicting the effects of their composition in dynamic social-computing environments. The model and simulation parameters can be changed dynamically, allowing quick testing of different incentive setups and behavioral responses at low cost. Specifically, we employ principles of agent-based *social simulation* [103, 61], an effective and inexpensive scientific method for investigating behavioral responses of large sets of human subjects. In theory, social simulation approaches (such as ours) allow modeling of incentives and responses of workers of arbitrary complexity. In practice, the social phenomena listed above as impeding factors for the development of comprehensive mathematical incentive models at the same time pose big obstacles for developing comprehensive simulation models, requiring development of complex agent behavioral models. Nonetheless, as discussed in [61, 178], the simulation approaches are a viable alternative to testing various behavioral responses in real communities when this is impossible due to time, cost or ethical reasons. All three limitations are especially accentuated when testing incentive effects and their different combinations. In this case, speed is preferred over accuracy and ethical considerations are an important feasibility factor. The simulation approach is therefore the method of choice in this case, offering fast experimental setups and circumventing ethical issues.

Social simulation originated in computational social science to explore theoretical ideas in the context of synthetic populations. Recently, this has been applied to crowd-sourcing, in order to generalize results which otherwise would be tied to a particular situation [22]. However, unlike the usual approach where agents interact directly (and thus benefit from cooperative behavior or suffer from defective behavior), we introduce a provider that facilitates interactions and determines the benefits or costs of those interactions.

8.3.3.1 Example Scenarios

In order to better describe the methodology and subsequently evaluate it, we first present two exemplifying scenarios based on real-world crowdsourcing applications.

Citizen-driven traffic reporting

Local governments have a responsibility to provide timely information on road travel conditions. This involves spending considerable resources on managing information sources as well as maintaining communication channels with the public. Encouraging citizens to share information on road damage, accidents, rockfalls or flooding reduces these costs while providing better geographical coverage and more up-to-date information[11]. Such a crowdsourcing process, however, poses data quality related challenges in terms of assessing data correctness, completeness, relevance and duplication.

Crowdsourced software testing

Traditional software testing is a lengthy and expensive process involving teams of dedicated engineers. Software companies[12] may decide to partially crowdsource this process to cut time and costs and increase the number and accuracy of detected defects. This involves letting remote testers detect bugs in different software modules and usage environments and submit bug reports. Testers with different reputations provide reports of varying quality and change the assigned bug severity. As single bugs can be reported multiple times in separate reports, testers can also declare two reports as duplicates.

The two scenarios exhibit great similarities. The expected savings in time and money can in both cases be outweighed by an incorrect setup and application of incentive mechanisms. Furthermore, the system could suffer from high numbers of

---

[11] For a real world example visit the Vienna City Council's "Sag's Wien" initiative at:
https://www.wien.gv.at/sagswien/

[12] For example, www.utest.com

purposely incorrect or inaccurate bug report submissions, driving the processing costs up. For the purpose of illustration, we join and generalize the two scenarios into a single, abstract one that we will use in our simulation setup.

The *Authority* seeks to lower the time and cost of processing a large number of *Reports* on various *Situations* occurring in the interest domain of the Authority. The *Workers* are independent agents, occasionally and irregularly engaging with the system managed by the Authority to perform one of the following *Actions*: *Submit* a new Report on a Situation, *Improve* an existing Report, *Rate* the accuracy and importance of an existing Report, inform the Authority that two existing Reports should be considered *Duplicates*. The Worker actions are driven by a combination of the following factors: a) the possibility to earn *Points* (translating into increased chances of exchanging them for money); b) the possibility to earn *Reputation* (translating into higher status in the community); and c) the intrinsic propensity of people to contribute and help or to behave maliciously. In order to influence and (de-)motivate workers, the Authority employs a number of *Incentive Mechanisms*, collectively referred to as the *Incentive Scheme*.

This scenario also needs to address the following challenges:

- *Crowdsourced report assessment.* The effort required for manual validation of worker-provided reports may easily outweigh the gained effort and cost reduction from crowdsourced reporting in the first place. Hence, workers need to be properly stimulated to supplement and enrich existing reports as well as vote on their importance, thereby lifting the verification burden off the Authority. The system also needs to strike a balance not to collect too much information.
- *Worker reputation (trust).* A worker's reputation serves as one potential indicator for data reliability, assuming that reputable workers are likely to provide mostly accurate information. Subsequently, reports from workers with unknown or low reputation need to undergo more thorough peer assessment. The system must support continuous adjustment of workers' reputation.
- *Adjustable and composable incentive scheme.* An effective incentive scheme needs to consider all past citizen actions, the current state of a report and the predicted costs of processing a report manually in order to decide whether and how to stimulate workers to provide additional information. It also needs to correctly identify and punish undesirable and selfish behavior (e.g., false information, deliberate duplication of reports, intentional up/downgrading of reports).

The resulting complexity arising from the possible combination and configuration of worker behavior, incentive schemes, and processing costs requires a detailed analysis to identify a stable and predictable system configuration and its boundaries.

8.3.3.2 Modeling and Simulation Methodology

Our methodology for simulating worker participation and incentive mechanisms in crowdsourcing processes is depicted in Figure 8.8. It consists of four basic steps, usually performed in multiple iterations:

1. defining a domain-specific meta-model by extending a core meta-model;
2. capturing worker's behavioral/participation patterns and reward calculation in an executable model;
3. defining scenarios, assumptions and configurations for individual simulation runs; and
4. evaluating and interpreting simulation results.

These steps are described in more detail below.

We use the DomainPro[13] modeling and simulation tool suite in each of the outlined methodology steps to design and instantiate executable models of incentive mechanisms and run simulations of those models. The tool allows creation of custom simulation languages through metamodeling and supports agent-based and discrete event simulation semantics (see [46]). However, the overall approach is generic and can be easily applied using a different modeling and simulation environment.

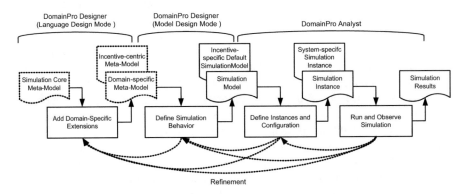

Fig. 8.8: The methodology of simulation design and development

The simulation core meta-model is implemented in the DomainPro Modeling Language. Optional extensions result in a domain-specific meta-model that defines which component types, connector types, configuration parameters and links a simulation model may exhibit. In our case, we extend the core meta-model to obtain what we refer to as an *incentive-centric meta-model*. The obtained incentive-centric meta-model serves as the basis for defining the simulation behavior, i.e., the *executable simulation model*. Obtaining the executable simulation model requires definitions of workers' behavioral parameters, Authority's business logic (including incentive mechanisms

---

[13] Tool available on request from:
www.quandarypeak.com

and cost metrics), the environment and the control flow conditions between them. Finally, prior to each execution, the executable simulation model requires a quick runtime configuration in terms of the number of Worker instances and monitored performance metrics (Section 8.3.3.3). During the execution, we do near real-time monitoring of metrics, and if necessary perform simulation stepping and premature termination of the simulation run to execute model refinements.

The tool we use enables refinement at any modeling phase. A designer will typically start with simple meta- and simulation models to explore the basic system behavior. She will subsequently refine the meta-model to add, for example, configuration parameters and extend the functionality at the modeling level. This enables simple incentive mechanisms to be tested first, and then extended and composed once their idiosyncrasies are well understood.

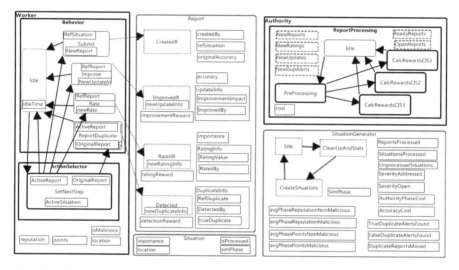

Fig. 8.9: Partial screenshot of the implemented case study simulation model in DomainPro Designer

Figure 8.9 provides a partial screenshot of the case study simulation model. The simulation model comprises over 40 simulation parameters, determining various factors, such as distribution of various personality characteristics in the Worker population, injected worker roles (e.g., malicious, lazy), base costs for the Authority, and selection and composition of incentive mechanisms. Describing them all in detail/formally here would not be practical. Instead, the annotated source code of the model and the meta-model is provided for download:[14]. The rest of this section is written in a narrative style.

Location and importance characterize a *Situation*. Situations can be generated with user-determined time, location and importance distributions, allowing us to

---

[14] http://tinyurl.com/incentives-sim-model

concentrate more problematic (important) situations around a predefined location in selected time intervals, if needed. For illustrative purposes, we generate situations with uniform probability across all three dimensions. The *SituationGenerator* contains the activities for creating new situations and calculating phase-specific simulation metrics on cost, reputation, points, actions and importance across reports, situations and Workers.

The *Worker's SetNextStep* activity represents the implementation of the worker's decision-making function f_a, introduced in Section 8.3.1. As previously explained, the Worker here considers the next action to perform based on:

1. internal state (e.g., *location*), including innate, population-distributed or arbitrarily set personality characteristics (e.g., *laziness*, *isMalicious*);
2. current performance metrics (e.g., *reputation*, *points*);
3. advertised rewards (*detectionReward*, *ratingReward*, *improvementReward*).

The Worker's location determines his/her proximity to a Situation, and, thus, the likelihood to detect or act upon that Situation (the smaller the distance, the higher the probability). However, two Workers at the same distance from a Situation are not equally likely to act upon it. This depends on their personality and past behavior, and the number of points they currently have. Default behavior of Workers is produced by normally distributing values of certain S state metrics, thus determining the "personality" of the Worker, in this case, the likelihood to act upon a situation. However, different Worker behaviors and personalities are obtainable through different *roles*. The simulation model does not prescribe the complexity of the roles. Instead, the incentive designer is free to implement them as necessary to simulate collective, disruptive of malicious behavior.

Points and *reputation* are the principal two metrics by which the Authority assesses Workers in our scenario. In principle, points are used by the Authority as the main factor to stimulate activity of a Worker. The more points, the less likely a Worker will idle. On the other hand, a higher reputation implies that the Worker will more likely produce artifacts of higher quality. Each new Worker joining the system starts with the same default point and reputation values. Precisely how the two metrics are interpreted and changed thereafter depends on the incentive mechanisms used (see below).

As we are primarily interested in investigating how reputation affects (malicious) behavior, we characterize each agent by a reputation metric, as laboratory experiments confirmed that reputation promotes desirable behavior in a variety of different experimental settings [184, 111, 145, 162].

The four *Behavior* activities produce the respective artifacts – *Reports*, *UpdateInfos*, *RatingInfos* and *DuplicateInfos*. The Worker's internal state determines the deviations in accuracy, importance, improvement effect, and rating value of the newly created artifacts. The subsequently triggered *Report*-located activities (*CreatedR*, *ImprovedR*, *RatedR* and *Detected*) determine the Worker action's effect on the two metrics that represent the artifact's state and data quality metrics at the same time – Report *accuracy* and *importance*. We use Bayes estimation to tackle the cold-start assessment of report accuracy and importance, taking into account average values of existing Reports and the reputation of the Worker himself.

The produced artifacts are queued at the *Authority* side for batch processing. In *PreProcessing* activity we determine whether a Report is ready for processing. This depends on the Report's quality metrics, which in turn depend on the number and value of Worker-provided inputs.

Processing Reports causes costs for the Authority. The primary cost factors are low-quality Reports and undetected duplicate Reports. Secondary costs arise when Workers focus their actions on unimportant Reports while ignoring more important ones. Therefore, the Authority incentivizes the Workers to submit required quantities of quality artifacts. As noted in Section 8.3.3.1, gathering as much inexpensive data from the crowd as possible was the reason for the introduction of a crowdsourced process in the first place.

Our proof-of-concept simulation model for the given scenario defines three basic incentive mechanisms:

- IM_1: Users are assigned a fixed number of points per action, independent of the artifact. Submitting yields most points.
- IM_2: The number of points is increased before assignment, depending on the current quality metrics of the Report. For example, the fewer ratings or improvements the higher the increment in points.
- IM_3: Users are assigned a reputation. The reputation rises with accurately submitted Reports, useful Report improvements, correctly rated importance and correctly flagged duplicates.

In Section 8.3.3.3, we compose these three mechanisms in different ways to produce different incentive schemes which we run and compare.

For demonstration purposes we define only a single additional role - that of a malicious worker. Malicious Worker behavior is designed to cause maximum cost for the Authority. To this end, we assume malicious Workers to have a good perception of the actual Authority characteristics. Hence, upon submission they will set initial Report importance low and provide very inaccurate information subsequently. For important existing Reports they will submit negative improvements (i.e., conflicting or irrelevant information) and rate them low, while doing the opposite for unimportant Reports.

8.3.3.3 Experimental Setup

Timing Aspects. We control the pace of the simulation by determining the number of Situations created per phase. Taking a reading of all relevant (i.e., experiment-specific) metrics at the end of each phase provides an insight on how these metrics change over time. All our simulations last for 250 time units (t), consisting of 10 phases of $25t$ each. Batch creation of Situations is representative for real world environments, such as bugs that typically emerge upon a major software release or spikes in traffic impediments coinciding with sudden weather changes. Report submission takes $5t$, while improving, rating and duplication flagging require only $1t$. The exact values are irrelevant as we only need to express the fact that reporting requires considerably more time than the other actions. Processing of Worker-provided data on the provider side occurs every $1t$. Note that for the purposes of the case study, here we are only interested in the generic processing costs rather than the time it takes to process that data. Each Report is assumed to cause 10 cost units for minimum quality (modeled as a value of 0), and almost no cost when quality (through Worker-provided improvements) approaches maximum ($= 1$). The higher the quality of received Reports, the fewer Reports are needed to persuade the Authority to act (see below).

Scenario-specific thresholds. As we aim for high-quality data and significant crowd-base confirmation, the following thresholds need to be met before a Report is considered for processing: at least three updates and high accuracy (> 0.75); or five ratings and medium importance (> 0.5); or four duplication alerts; or being reported by a Worker of high reputation (> 0.8) and having high importance (> 0.7). Workers obtain various numbers of points for (correct) actions, the number depending on the value of the action to the provider and the incentive scheme used.

Worker Behavior Configuration. A Worker's base behavior is defined as 70% probability idling for $1t$, 20% submitting or duplication reporting, and 10% rating or improving. Obtained points and reputation increase the likelihood to engage in an action rather than idle. The base behavior represents rather active Workers. We deliberately simulate only the top k most involved Workers in a community as these have the most impact on benefits as well as on costs. Unless noted otherwise, $k = 100$ for all experiments.

Composite Incentive Schemes. The experiments utilize one or more of the following three previously introduced *Composite Incentive Schemes – CIS*:

- $CIS1 = IM_1$
- $CIS2 = IM_2 \circ IM_1 = IM_2(IM_1)$
- $CIS3 = CIS2 + IM_3 = IM_2 \circ IM_1 + IM_3$

CIS1 promises and pays a stable number of points for all actions. CIS2 dynamically adjusts assigned points based on the currently available Worker-provided data, but at least as high rewards as CIS1. CIS3 additionally introduces reputation calculation.

8.3.3.4 Experiments

Experiment 1: Comparing Composite Incentive Schemes.
Here we compare the impact of CIS1, CIS2 and CIS3 on costs, assigned rewards, Report accuracy and timely processing. Figure 8.10 displays incurred costs across the simulation duration. All three schemes prove suitable as they allow 100 Workers to provide sufficient data to have 20 Situations processed at equally high accuracy. They

differ significantly, however, in cost development (Fig.8.10 inset), primarily caused by undetected duplicate Reports (on average 0.2, 0.25 and 0.4 duplicates per Report per phase for CIS1, CIS2 and CIS3, respectively). CIS1 yields stable and overall lowest costs as the points paid induce just the right level of activity to avoid Workers getting too active and thus causing duplicates. This is exactly the shortcoming of CIS2 which overpays Workers who subsequently become overly active. CIS3 pays even more, and additionally encourages Worker activity through reputation. The cost fluctuations are caused by the unpredictable number of duplicates (however remaining within bounds). Although more costly and less stable, CIS3 is able to identify and subsequently mitigate malicious Workers (see Experiment 3 below).

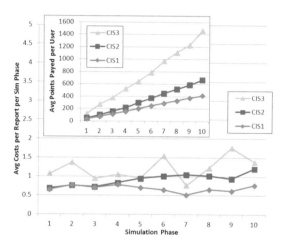

Fig. 8.10: Incurred Report-processing costs for CIS1, CIS2 and CIS3. Inset: average points paid per Worker

Experiment 2: the Effect of Worker/Situation mismatch.
Here we analyze the effects of having too few or too many Workers per Situation. In particular, we observe per phase the cost, points assigned, Report importance (as reflecting Situation importance) and reputation when:

1. the active core community shrinks to 20 Workers while encountering 50 Situations (20u/50s);
2. a balance of Workers and Situations (100u/25s);
3. many active Workers but only a few Situations (100u/5s).

A surplus of Situations (20u/50s) causes Workers to become highly engaged, resulting in rapid reputation rise (Fig 8.12 bottom) coupled with extremely high values of accumulated rewarding points (Fig 8.11 inset). Costs per Report remain low as duplicates become less likely with many Situations to select from (0.18 duplicates per Report). Here, CIS3 promises more reward for already highly rated Reports to counteract the expected inability to obtain sufficient Worker input for all Situation (on average 22 Reports per phase out of 50). Subsequently, the Authority receives correct ratings for Reports and can focus on processing the most important ones. Compare the importance of addressed Situations in Figure 8.12 top. A surplus of active Workers (100u/5s) suffers from the inverse effect. As there is little to do, reputation and rewards grow very slowly. Perceiving little benefit, Workers may potentially leave while the Authority has a difficult time distinguishing between malicious and non-malicious Workers. Configurations (100u/5s) and (100u/25s) manage to provide Reports for all Situations, therefore having average Report importance remaining near 0.5, the average importance assigned across Situations.

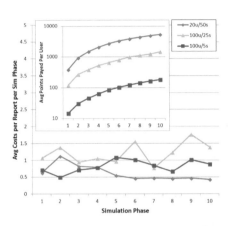

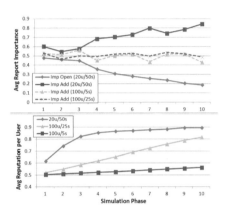

Fig. 8.12: Reputation acquired by Workers (bottom), and Report **impo**rtance **add**ressed, respectively remaining **open** (top)

Fig. 8.11: Costs per Report incurred at various combinations of Worker and Situation count.

Experiment 3: Effect of Malicious Workers.
Here we evaluate the effects of an increasing number of malicious Workers on cost when applying CIS3. Figures 8.13 and 8.14 detail cost and reputation for 0%, 20%, 30%, 40% and 50% malicious Workers. All Workers are considered of equal, medium reputation 0.5 upon simulation start. The drop in costs across time (observed for all configurations) highlights that the mechanism indeed learns to distinguish between regular, trustworthy Workers and malicious Workers. The irregular occurrence of undetected duplicates causes the fluctuations in cost apparent for 0% and 20% malicious Workers. Beyond that, however, costs are primarily determined by low accuracy induced by malicious Workers. CIS3 appears to work acceptably well up to 20% malicious Workers. Beyond this threshold harsher reputation penalties and Worker blocking (when dropping below a certain reputation value) need to be put in place. In severe cases lowering the default reputation assessment might be applicable

but requires consideration of side effects (i.e., thereby increasing the entry barrier for new Workers).

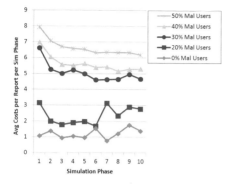 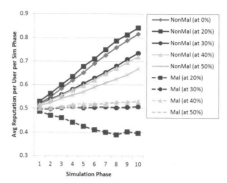

Fig. 8.13: Costs per Report incurred due to various levels of malicious Workers

Fig. 8.14: Average reputation acquired by malicious and non-malicious Workers

8.3.3.5 Limitations and Discussion

Simulations of complex socio-technical processes such as the use case presented here can only cover particular aspects of interest, never all details. Thus any results in terms of absolute numbers are unsuitable for direct application in real-world systems. Instead, the simulation enables incentive scheme engineers to compare the impact of different design decisions and decide what trade-offs need to be made. The simulation outcome provides an understanding which mechanisms might fail earlier, which strategies behave more predictably and which configurations result in a more robust system design.

In particular, the presented comparison of CISs in Experiment 1 gives insight into the impact of overpaying as well as indicating that CIS3 would do well to additionally include a mechanism to limit submissions and better reward the action of flagging duplicates. Experiment 2 provides insights on the effect of having too few or too many Workers for a given number of Situations. It highlights the need to adjust rewards and reputation in reaction to shifts in the environment and/or worker community structure. Experiment 3 provides insight into the cost development in the presence of malicious workers and highlights the potential for mechanism extension.

8.4 PRINGL – A Programming Framework for Incentive Management

In the previous section we have introduced the Rewarding Model (RMod), which is capable of encoding incentives in the form of scheduled application of rewarding actions applied over an abstract model. We then showed through simulation that the model was capable of detecting individuals with dysfunctional behavior and reducing their influence. In this chapter we present the design of a programming framework making use of the introduced RMod and allowing application of rewarding actions on real-world socio-technical systems.

The incentive management programming framework is intended to be used by two types of *users*:

1. *incentive designers* – domain experts who design and implement incentive scheme for an organization;
2. *incentive operators* – organization members responsible for managing the every-day running and adaptation of the scheme.

An incentive designer (*Designer*) is a multidisciplinary domain expert in areas spanning management, economy, game theory and psychology. The Designer designs on behalf of the Smart City platform a set of appropriate incentive mechanisms for the given business model of the value-added application, taking into consideration context-specific properties pertinent to the targeted population of workers/citizens. An example of how this process is performed for two different experimental platforms can be found in [51, 3].

The role of an incentive operator (*Operator*) has not been defined in the existing literature, as its existence is subject to the existence of the novel type of incentive management platforms that we describe here. While a Designer can be a person external to the socio-technical platform, the Operator is a member of the management of the socio-technical platform in charge of monitoring the application of incentives and taking operative decisions on adaptations of various incentive parameters.

To exemplify the expected type of functionality an Operator performs, let us assume the existence of a socio-technical platform offering a crowdsourced software development service to its customers. The Operator's role is to monitor the efficacy of incentive schemes in use and adjust them when needed. For example, the operator might learn that teams in which testers were incentivized to report more bugs throughout the entire development process performed worse than those incentivized to report (fewer, but) more severe/dangerous ones in more mature product phases. Based on this experience, the Operator can adjust the scheme (e.g., bug thresholds and bonus amounts) to put more emphasis on quality rather than on quantity as soon as the product has entered a fairly stable stage.

Both the Designer and the Operator can use the simulation modeling methodology introduced in Section 8.3.3 to aid the design, composition and adjustment of the incentive scheme. Operators in particular can benefit from the speed that the social simulation offers (compared to the conventional incentive mechanism design) when adaptations of the incentive scheme (e.g., parameter variations, turning on/off additional mechanisms) are necessary to counteract disruptive or newly emerging dysfunctional behavior. Both Designer and Operator use a platform-independent, largely declarative domain-specific language to encode/adjust incentive schemes that are provided as input to the incentive management platform.

In this section we present the framework's programming and execution model, the semantics and syntax of PRINGL[15]. We describe PRINGL's modeling paradigm, and demonstrate its expressiveness by modeling a set of realistic incentive mechanisms.

In order to enact a PRINGL-encoded incentive on a socio-technical platform (i.e., apply the incentives on real crowd workers), we need a simplified and uniform model of the platform's workers, and the metrics and relationships that describe them. We call such a model together with the framework that manages it an *abstraction interlayer* (Fig. 8.15). More precisely, we use the term abstraction interlayer to denote any middleware sitting on top of a socio-technical system, exposing to external users a simplified model of its employed workforce and allowing monitoring of the workers' performance metrics. The existence of an abstraction interlayer allows the incentive designer to write fully portable incentives.

The PRINC framework (introduced in Section 8.3.2) possesses all the characteristics of an abstraction interlayer. It features the general model (RMod) for representing the state of a socio-technical system, reflecting its quantitative, temporal and structural aspects. PRINC's mapping model (MMod) defines the mappings needed to properly express the platform-specific versions of metrics, actions, artifacts and attributes in their RMod cognates. Finally, PRINC coupled with SMARTCOM takes care of exchanging messages with and receiving update events from the underlying socio-technical platform, thus enabling the RMod abstract model to mirror the state of the underlying system. This in turn allows us to express incentive mechanisms decoupled from the underlying platform: to apply an incentive it suffices to alter the RMod state, while the task of mirroring this change on the actual socio-technical platform is delegated to PRINC and SMARTCOM.

In this chapter we assume the existence of PRINC as abstraction interlayer. The business logic code provided in the examples in Section 8.4.4 is C# code executable on PRINC. In theory, PRINGL can work without an abstraction interlayer. However, this would imply that all message handling with the underlying crowdsourcing system and complex monitoring logic would have to be written from scratch. This contradicts one of the principal motives for the introduction of PRINGL, and is more disadvantageous than building a completely system-specific incentive management solution.

8.4.1 Overview

Figure 8.15 shows an overview of PRINGL's architecture and usage in the overall context of the incentive management framework. An incentive designer models an

---

[15] PRogrammable INcentives Graphical Language– a domain-specific language for modeling incentives for socio-technical systems.

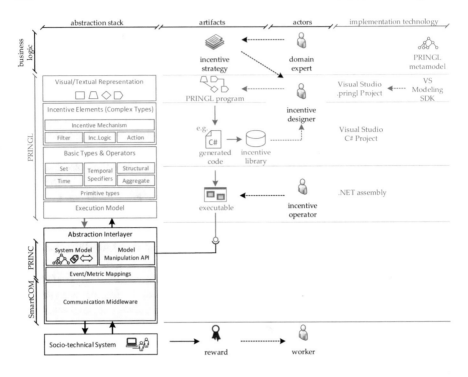

Fig. 8.15: Incentive management framework tools, showing an overview of PRINGL's programing model elements, architecture, users, operative environment and implementation.

incentive scheme provided in natural language by a domain expert as a PRINGL program using PRINGL's visuo-textual syntax. The visually expressed part of the syntax is completely system-independent, while system-specific business logic can be expressed as source code in an arbitrary programming language supported by the abstraction interlayer. In this respect, its portability equals the portability offered by the abstraction interlayer.

Starting from a PRINGL program the PRINGL code generator produces the following artifacts, encoded in a conventional programming language:

- An incentive model expressed in terms of incentive elements, basic PRINGL types and operators. This model also integrates the business logic code provided by the incentive designer. The incentive element definitions from this model can optionally be compiled into libraries for later reuse.
- Code for communication with the abstraction interlayer and application of the incentives.
- Code for manipulation of the incentive model.

These artifacts can be used to quickly build applications offering incentive management capabilities, e.g., a GUI-based application offering an incentive operator the possibility to change the runtime parameters. As previously explained, the abstraction interlayer takes care of communication with the concrete socio-technical system, forwarding the rewarding actions and receiving the updates. To do this, we make use of the SMARTCOM middleware introduced in Section 7.2. The abstraction interlayer delegates the burdensome tasks of communication and privacy management to SMARTCOM, allowing advertising of incentives and monitoring of worker participation over popular (commercial) communication protocols. Figure 8.16 shows the intended usage of SMARTCOM in the incentive management framework context.

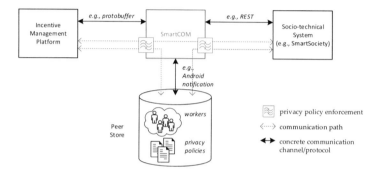

Fig. 8.16: SMARTCOM's application context

8.4.1.1 Requirements

As PRINGL is a domain-specific language, the focus of the design requirements lies primarily on coverage of the domain and usability by the stakeholders within that domain. The design of the language was guided by the following requirements, formulated according to the guidelines outlined in [113]:

1. *Usability* – provide an intuitive, user-friendly modeling DSL for incentive operators.
2. *Expressiveness* – provide an expressive environment for programming complex real-world incentive strategies for incentive designers.
3. *Groundedness* – allow the use of *de facto* established terminology, components and methods for setting up incentive strategies.
4. *Reusability* – support and promote reuse of existing incentive business logic.
5. *Portability* – support system-independent incentive mechanisms, agnostic of type of labor or workers, and of underlying systems.

8.4.2 Programming Model

To meet the specified requirements PRINGL was conceived as a hybrid visual/textual programming language, where incentive designers can encode core *incentive elements*, while incentive operators can provide concrete runtime parameters to adapt them to a particular situation. The language supports programming of the real-world incentive elements described in [155, 172] and allows composition of complex *incentive schemes* out of simpler elements. Such a modular design also promotes reusability since the same incentive elements with different parameters can be used for a class of similar problems, stored in libraries and shared across platforms.

PRINGL allows incentive designers to model natural-language, realistic incentive schemes (i.e., business logic) into a platform-independent specification through a number of incentive elements represented by a visual syntax (graphical elements with code snippets). The incentive scheme represents the whole business logic needed for managing incentives in an organization. The scheme is expressed in PRINGL as a number of prioritized *incentive mechanisms* representing a PRINGL program. Each mechanism can then be further decomposed into a number of constituent incentive elements described in the following subsections. The designer programs new incentive elements or reuses existing ones from an incentive library to compose new, more complex ones. The following sections describe the incentive elements and operations on them.

8.4.2.1 Primitive Incentive Elements

From a business logic perspective, primitive incentive elements represent the basic entities (workers, relationships and time units) that we use when composing incentive rules. From a programming language perspective, they can be considered as atomic types that are used in user-provided or library code that specifies business logic. We use the two terms "type" and "incentive element" interchangeably. Apart from the four conventional primitive types: `string`, `bool`, `int` and `double`, PRINGL defines the types shown in Table 7. They do not have a direct visual representation. Only primitive elements can be used as inputs and outputs of *complex incentive elements* (Section 8.4.2.3).

PRINGL provides a number of operators for manipulating the introduced primitive types.

8.4.2.2 Built-In Operators

- *Set operators.* – Union, intersection and complement on `Collection<T>`.
- *Time operators.* If working with adjustable intervals, it is advisable to use operators wherever possible as they are evaluated at runtime and guarantee that any external changes (e.g., deadline extensions) will be taken into account. A common use case sees a user initializing an `Interval` from an iteration, and

Type	Description
Worker	Represents an individual worker and his/her performance metrics
PoiT	Represents a point in time. It can be instantiated by providing a fixed datetime or obtained as the result of application of time operators
Interval	Represents a named, addressable time interval. An interval can be: a) *fixed*; or b) adjustable. Fixed intervals have predefined starting and ending times, provided by two PoiTs, that cannot subsequently be altered. Adjustable intervals reflect the external system's changes to intervals, e.g., deadline extensions (cf. *iterations* [156]). Changes are allowed to affect only points in the future
Collection<T>	An iterable collection of a primitive type T is also considered a primitive type

Table 7: Primitive PRINGL types.

using interval operators to specify points in time at which an action is needed. Time operators are commonly used with temporal specifiers.

- StartOf(Interval i) – returning the Collection<PoiT> containing a single time point representing the interval's currently expected starting time.
- EndOf(Interval i) – returning the Collection<PoiT> containing a single time point representing the interval's currently expected ending time.
- PartOf(Interval I, double p) – $p[0,1]$ returning the PoiT at percentage p of the interval. $PartOf(i,0) == StartOf(i); PartOf(i,1) == EndOf(i)$
- MultiPoint(Interval i, int k) – returns a Collection<PoiT> of points evenly distributed between StartOf() and EndOf().
- AllOf(Interval i) – returns a Collection<PoiT> of points representing all time points (depending on the resolution of the underlying system) contained in the interval.

- *Temporal specifiers.* These are special operators used to instruct the execution environment when to perform certain actions or evaluate predicates. As such, they cannot be directly used in user-provided programming code, but are rather offered as a choice through a visual GUI element (drop-down box) where needed. Internally, they are represented as built-in functions that operate on a collection of PoiTs that is provided by the environment at runtime.

  - Always(Collection<PoiT>) – "at each PoiT in collection".
  - Sometimes(Collection<PoiT>) – "at least once in collection".
  - Once(Collection<PoiT>) – "exactly once in collection".
  - Never(Collection<PoiT>) – "never in collection".
  - First(Collection<PoiT>) – "oldest in collection".
  - Last(Collection<PoiT>) – "newest in collection".

- *Structural operators.* These perform structural queries/modifications by exa-
  mining/re-chaining the relationships between worker nodes in the abstraction
  interlayer (graph) model by using *graph transformations* [14].

    - Querying:
        · `neighborsOf(Worker w, string relationType, int numHops,`
          `bool directed)` – returns a `Collection<Worker>` filled with work-
          ers `numHops` hops away from `Worker w` over un-/directed `relationType`
          relationships.
        · `managersOf(Worker w)` – returns `Collection<Worker>` filled with
          manager(s) of worker W. The relationship type representing the man-
          agerial relation is obtained from the abstration interlayer.
        · `subordinatesOf(Worker w)` – analogous to `managersOf`.
    - Modifying:
        · `changeManager(Worker w, string teamLabel)` – re-chains the
          implicitly determined managerial relations within the members of the
          `teamLabel` team to point to the new manager.

- *Aggregation operators.* These perform calculations on performance metrics
  or events over a `Collection<PoiT>`, in a fashion similar to SQL's aggregate
  functions. The collection of time points over which the operators calculate is
  provided by the runtime environment at each invocation. They can only be used
  in predicate logic blocks ⟨P⟩ that are directly or indirectly reachable through
  declaration relationships originating from a `WorkerFilter` ⟨F⟩ element.

    - `@AVG(double` *m*`)` – returns the average value of the metric *m* over the given
      time point collection.
    - `@COUNT(string` *evt*`)` – returns the number of occurrences of event *evt* in the
      timespan delimited by the first and last `PoiT` in the given input collection.
    - `@MAX(double` *m*`)` – returns the largest value of the metric *m* over the given
      time point collection.
    - `@MIN(double` *m*`)` – returns the smallest value of the metric *m* over the given
      time point collection.
    - `@SUM(double` *m*`)` – returns the sum of the values of the metric *m* over the
      given time point collection.

8.4.2.3 Complex Incentive Elements

Complex types enable PRINGL's core functionality and are represented by corresponding graphical elements. Their key property is that more complex types can
be obtained by visually combining simpler ones. Visual, rather than purely textual
representation was chosen to allow users to build up complex incentive schemes
by visually suggesting and restricting the choice of the possible components, thus
facilitating the process of construction of incentive mechanisms. Complex incentive
elements are managed through the following operations.

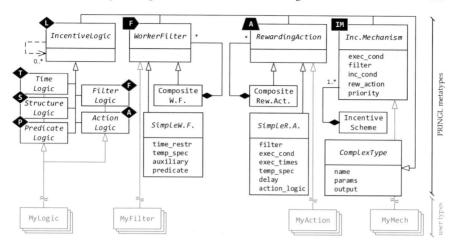

Fig. 8.17: Complex incentive elements class hierarchy

8.4.2.4 Operations on Complex Incentive Elements

Definition – Complex types are defined by inheriting the following abstract metatypes: IncentiveLogic, WorkerFilter, RewardingAction and IncentiveMechanism (Fig 8.17). A new complex type inherits the predefined, addressable *fields* from the metatype it redefines. In order for a type definition to be complete, the fields must be filled out with appropriate values. Some fields are filled out automatically by PRINGL depending on the context where they are used (auto parameters); others must be filled out by the user (user-fields). The user-fields are: a) name, which specifies the name of the new complex type; b) an arbitrary number of primitive-type input parameters (params) that can be used in evaluations and passed to other incentive elements; c) type-specific fields, specifying how a particular functionality of the newly defined complex type is going to be executed – by indicating another incentive element to invoke, or by providing an executable code snippet. Definition is performed through appropriate graphical constructs being placed onto the working area. A new type definition retains its parent metatype's graphical representation. For the non-auto input params (b), PRINGL visually exposes the appropriate number of GUI form fields accepting the inputs that are to be filled out manually by the user. The input can contain expressions with primitive types and/or references to other accessible fields. To fill out type-specific fields (c), the user is expected to visually link the appropriate incentive element type, thus effectively declaring/instantiating it (see below).

Declaration/Instantiation – When defining new complex types, the user indicates (declares) which field/subcomponent instances will be required at PRINGL runtime to instantiate the newly defined object by placing the corresponding graphical (filled) element in the appropriate context within the working area, connecting it with appropriate connectors from the parent type definition, and overriding parameter values from the parent type definition, if needed. The auto parameters are loaded at instantiation by PRINGL transparently to the user. For example, in case of ⟨T⟩ (Section 8.4.2.5) all named Intervals and all workers are passed as input parameters and made available through predefined variable names (preceded with underscore). This removes the need to know how to access certain data from a type definition, thus making it self-contained and portable. The user-defined fields are initialized with values calculated from the expression contained in the type definition and values provided by the user or propagated from the composing elements. Type instances are addressable objects that can be referenced (e.g., to read a field value) or invoked (see below) from the programming code and other elements.

Indirect invocation – The IncentiveLogic, WorkerFilter and RewardingAction instances can also be "invoked" just by being referenced from expressions in user code. When the PRINGL code generator encounters an instance reference in an expression it transparently replaces it with an invocation of the default method for that type. Default methods for filters and rewarding actions return the resulting Collection<Worker>. The *default method* of an IncentiveLogic type is a function having input and output parameters as specified in its definition, and the user-provided code as the function body. The input parameters are provided by PRINGL runtime, so there is no need to pass any non-user parameters from the user code. Expressions containing indirect invocations can be used as field values (see Ex. 8.4.4.2, Fig. 8.27) or arbitrarily within the user-provided business logic code in IncentiveLogic elements (see Ex. 8.4.4.3, Fig. 8.25, ①). The indirect invocation feature allows the user to pass instance references instead of output types of their default methods; for example, we can pass a filter instance to an IncentiveLogic element expecting a single input parameter of type Collection<Worker>. As these are common situations, indirect invocation helps cut down on verbosity of user code.

Static invocation – In addition to indirect invocation, IncentiveLogic elements can be invoked statically with arbitrary input parameters from the user code. In order to make the static invocation, the IncentiveLogic type name is appended with .invokeWith([<*param-list*>]); see Ex. 8.4.4.3, Fig. 8.25, ①.

8.4.2.5 Defining Complex Incentive Elements

Incentive Logic ◇

These constructs encapsulate different aspects of business logic related to incentives in reusable fragments (e.g., determine whether a condition holds, read a metric value, or perform a simple action). They can be thought of as functions/delegates with predefined signatures allowing only certain input and output parameters. They are invoked from other PRINGL constructs, including other IncentiveLogic elements. Implementation is dependent on the abstraction interlayer, but not necessarily on the underlying socio-technical platform, meaning that many libraries can be shared across different platforms, promoting reusability of proven incentives, uniformity and reputation transfer. The Designer is encouraged to implement incentive logic elements as small code snippets with intuitive and reusable functionality. Depending on the intended usage, incentive logic elements have different subtypes: Action, Structural, Temporal, Predicate, Filter. Subtypes are needed to impose necessary semantic restrictions, e.g., the subtype prescribes different input parameters and allows PRINGL to populate some of them automatically[16]. Similarly, different subtypes dictate different return value types. These features encourage high modularization and uniformity of incentive logic elements. Descriptions of the incentive logic subtypes are provided in Table 8. The incentive logic element definition is expressed in PRINGL with the visual syntax element shown in a Fig. 8.18, with the appropriate subtype symbol shown in the upper left corner. As is the case with other incentive element definitions (presented in subsequent sections), the incentive logic element incorporates the distinguishing geometrical shape (diamond in this case), as well as auto-populated and user-defined parameters. Differently than other elements, it contains a field into which the Designer inputs executable code in a conventional programming language. The code captures the business logic specific to the incentive that is being modeled, but must conform to the rules imposed by the incentive logic subtype. As a shorthand, textual, inline notation for incentive logic elements we use a diamond shape surrounding the letter indicating the subtype, e.g., ⟨T⟩ for temporal logic.

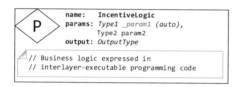

Fig. 8.18: Visual element representing an `IncentiveLogic` definition

---

[16] Marked with `auto` in figures

Subtype	Symbol	Environment-provided input	Allowed output	Intended usage
TimeLogic	⊤	all named Intervals, all Workers, reference to global state	Collection <PoiT>	To return time intervals/points at which a predicate should be evaluated or an action performed
StructureLogic	S	reference to the structural model, reference to global state	Collection <Worker> for queries: found workers; for transformations: affected ones	To perform graph queries/transformations on the model representing workforce structure and relationships. A transformation S is only allowed to be invoked from A. A query S can only be invoked from P and F
PredicateLogic	P	currently evaluated Worker, all Workers, currently evaluated PoiT, reference to global state	bool	To evaluate whether a predicate holds at a given moment
FilterLogic	F	currently evaluated Interval, all named Intervals, currently evaluated Worker, all Workers, reference to global state	arbitrary	To provide business logic for evaluating past worker performance
ActionLogic	A	Workers to be rewarded/punished, reference to global state	Collection <Worker> (affected)	To perform rewarding actions over workers or global variables

Table 8: IncentiveLogic subtypes

Worker Filter ▷

The function of a `WorkerFilter` element is to identify, evaluate and return matching workers for subsequent processing based on user-specified criteria. The criteria are most commonly related (but not limited) to worker's past performance and team structure. The workers are matched across different time points from the input collection of `Workers` that is provided by the PRINGL environment at runtime. By default, all the workers in the system are considered. The output is a collection of workers satisfying the filter's predicate.

If we denote the input set of `Workers` of a `WorkerFilter` ▷ with I_x, and the output set with O_x, then the functionality of ▷ can be defined as the function f_x:

$$f_x : I_x \rightarrow O_x$$
$$I_x = input(x)$$
$$O_x = \{e \in O_x \mid e \in I_x \wedge p_x(e) = true\}$$

where p_x is the filter's predicate. Therefore, the functionality of a filter is to return a subset of workers from the input set, i.e., to perform a set restriction. Both `SimpleWorkerFilter` and `CompositeWorkerFilter` are subtypes of the abstract metatype `WorkerFilter` (Fig. 8.17), and can be used interchangeably where a worker filter is needed. A `SimpleWorkerFilter` element definition is expressed in PRINGL with the visual syntax element shown in Fig. 8.19, while a right-pointed shape ▷ is used as the inline, shorthand, textual denotation. A filter's type-specific fields are filled out visually by the user, by connecting them with appropriate incentive elements. Field descriptions are provided in Table 9.

Field	Description
`time_restr`	An optional ◁ returning a collection of time points which should be considered when evaluating workers. If omitted, the default value is a collection containing only a single `PoiT` representing the present moment
`temp_spec`	An optional temporal specifier (Section 8.4.2.1) determining how to interpret the filter predicate values across different time points. If unspecified, the predicate is evaluated only for the last (most recent) `PoiT` in the collection
`auxiliary`	An optional ◁ that is used to fetch some global metrics needed for worker evaluation, and possibly provide some intermediate results to be used for evaluating the filter predicate
`predicate`	A required ◁ providing the predicate that will be evaluated against each worker at specified time points

Table 9: `SimpleWorkerFilter` fields

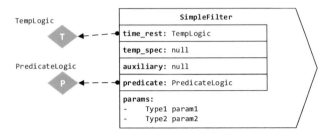

Fig. 8.19: Visual element used for `SimpleWorkerFilter` definition

Composite Filters

In Figure 8.20 we illustrate how a composite filter can be defined in PRINGL. It consists of graphical elements representing instances of previously defined or library-provided `WorkerFilters`. The elements are connected with directed edges denoting the flow of `Workers`. There must be exactly one filter element without input edges representing the *initial filter*, and exactly one filter element without output edges representing the *final filter* in a composite filter definition. When a `CompositeWorkerFilter` is instantiated and executed, PRINGL provides the input for the initial filter, and returns the output of the final filter as the overall output of the composite filter. Like any other PRINGL composite type, a composite filter can also expose propagated or user-defined parameters.

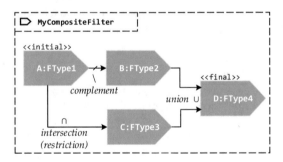

Fig. 8.20: An example `CompositeWorkerFilter` definition

A directed edge $\mathbb{A} \rightarrow \mathbb{C}$ implies that $\mathbb{C}$ takes as input $\mathbb{A}$'s output (the workers matching the criteria of $\mathbb{A}$). The output of $\mathbb{C}$ is a set containing workers fulfilling both filters' conditions, thus effectively representing the $\mathbb{A} \cap \mathbb{C}$ operation. If an edge is marked as *negating* ($\nrightarrow$), then $\mathbb{A} \nrightarrow$ returns the set complement of $\mathbb{A}$'s input, i.e., $input(A) \setminus \mathbb{A}$. When multiple edges enter a single filter element, then the union ($\cup$) of workers coming over the edges is used as the input for the filter element. When multiple edges go out of a single element, then the same output set of workers is passed to each receiving end. Sometimes, we need a filter to forward the same set of workers to multiple filters or to collect workers from multiple filters without performing additional restrictions; the *pass-through* filter (predefined `PassThru` type) contains no logic, except for a predicate always returning `true`.

Rewarding Action

Its function is to notify the abstraction interlayer (and consequently the socio-technical platform) that a concrete action should be taken regarding specific workers at a given time, or that certain specific actions should be forbidden to some workers during a certain time interval. The rewarding actions can include, but are not limited to, the following: adjust reward rates (e.g., salary, bonus), assign digital rewards (e.g., points, badges, stars), suggest promotion/demotion or team restructuring, display a selected view of rankings to selected workers. The choice of the available actions is dependent on the set supported by the interlayer and the actual socio-technical platform. The abstraction interlayer is responsible for translating the action into a system-specific message and delivering it to the underlying platform. PRINGL expects the underlying system to acknowledge via the abstraction interlayer that the suggested action was accepted and applied to a worker, because its outcome may affect other incentive mechanisms. We use the trapezoid shape shown in Fig. 8.21 to denote the definition of a `SimpleRewardingAction`. For the shorthand notation, we use ⟋A⟍, both for simple and composite rewarding action elements.

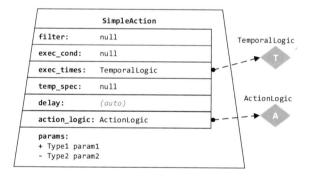

Fig. 8.21: Visual element used for `SimpleRewardingAction` definition

In PRINGL's programming model the output of a `RewardingAction` is a `Collection<Worker>` containing affected workers, i.e., those to whom the action was successfully applied. Informing the abstraction layer is performed as a side-effect of executing the rewarding action. In order to perform the action, the runtime environment needs to know to which workers the action applies, so a worker filter needs to be used (`filter` field). In some cases, the workers that are rewarded/punished may be the same as the initially evaluated ones. In that case we can reuse the original filter used for evaluation. In other cases, workers may be rewarded

Field	Description
filter	An optional $\langle\mathbb{F}\rangle$ determining the workers to whom to apply the action. If omitted, the worker collection is by default provided by the runtime environment from the output of the original evaluation filters
exec_cond	An optional $\langle\mathbb{P}\rangle$ establishing whether the currently evaluated worker earned the reward/punishment or not. If omitted, considered 'true' by default
exec_times	An optional $\langle\mathbb{T}\rangle$ returning Collection<PoiT> determining the possible execution points. If omitted, the environment defaults to the current PoiT and executes immediately
temp_spec	An optional temporal specifier further restricting the original collection of execution PoiTs. Defaults to Always() if omitted
delay	A hidden parameter set by the environment and used for recalculating execution times in composite rewarding actions. It contains a non-negative integer time offset added to the execution PoiTs. The actual time unit is determined as the basic time unit of the underlying layer (an RMod *tick* in our case). The default value is zero
action_logic	A mandatory reference to an $\langle\mathbb{A}\rangle$ element containing the system-specific business logic that invokes the rewarding action

Table 10: SimpleRewardingAction fields

based on the outcome of evaluation of other workers (e.g., team managers for the performance of team members). PRINGL's runtime also needs to determine the timing for action application (temp_spec and exec_times fields). We use temporal speci-fiers (see Sec. 8.4.2.1) to determine the exact time moment(s) of the time series. For defining incentives involving *deferred compensation* [155] we also need to specify an additional predicate that will be evaluated at the execution time to establish whether a worker fulfilled the reward criteria during the period from when the incentive was scheduled until the execution point (exec_cond field). The actual action to execute is determined by the action_logic field, pointing to a concrete $\langle\mathbb{A}\rangle$ element. To execute the action PRINGL needs to invoke the appropriate action in the abstraction interlayer, which will then send out a system-specific message to the underlying platform. Field descriptions are summarized in Table 10.

Composite Actions

Similarly to composite filters, a CompositeRewardingAction definition consists of graphical elements representing instances of previously defined RewardingActions. It must contain exactly one *initial action* a_0, and exactly k_0 *final actions*, where k_0 is the number of a_0's outgoing edges. The elements are connected with directed edges denoting at the same time: a) *Worker flow*; and b) *time delay*. There must be no cycles in the graph, i.e., the flow must be a tree with the root in the initial action, with each final action being a leaf. Like any other PRINGL composite type, a composite action can also expose propagated or user-defined parameters.

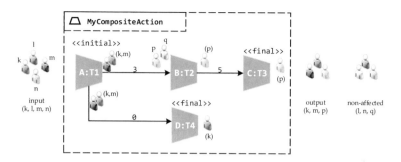

Fig. 8.22: An example `CompositeRewardingAction` definition with branch delays shown

Worker flow. A `RewardingAction` returns *affected* workers and passes them over outgoing edges. Affected workers are those workers to whom the action was successfully applied by the underlying system. The definition of a successful application is system-specific. Therefore PRINGL expects the underlying system to acknowledge via the abstraction interlayer that the suggested action was accepted and successfully applied to the worker. The passing of workers is similar to that of composite filters. The two major differences are:

1. The absence of graph cycles prevents the union ($\cup$) operation on passed worker sets.
2. Any `RewardingAction` element can either use the provided input workers, or completely ignore them, and identify the input workers by itself. For example, a `SimpleRewardingAction` does it by initiating the optional `filter` field. This limitation allows the worker flow to be changed at arbitrary places in the composition.

Figure 8.22 shows an example of a `CompositeRewardingAction` definition. It also shows an example of worker passing. The initial action ⟋A⟍ is given the set (k,l,m,n) as input. The execution of ⟋A⟍ ends with successful rewarding of workers (k,m). This intermediate set is immediately added to the resulting output set. The same intermediate set of workers is passed to actions ⟋B⟍ and ⟋D⟍. Action ⟋D⟍ ends with rewarding only one of those workers – (k). k is already part of the output, so nothing else happens on this execution branch. The action ⟋B⟍, on the other hand, discards the input worker set (k,m), and determines its own input set (p,q). After execution, ⟋B⟍ returns just (p), which is also added to the aggregate output set and passed further as input to ⟋C⟍, which also happens to reward p successfully.

Time delay. Each edge can optionally specify a time delay as a non-negative integer without the unit. If omitted, zero is assumed. The actual unit is determined transparently to the user as the basic time unit of the abstraction interlayer. PRINGL forwards the delay value to the action to which the edge points.

If this action is a `SimpleRewardingAction`, this equals to adding the specified time offset to the hidden `delay` parameter. Later, when executing the action, PRINGL will add the value of the `delay` parameter to each PoiT returned by the action's `exec_times` ⟨↑⟩. If the delay is forwarded to a `CompositeRewardingAction`, then the delay is forwarded to its initial action.

The execution of a composite action starts by first breaking it into linear execution paths containing constituent simple actions. For each execution path PRINGL then takes into account specified delays for each simple action and immediately schedules it with the abstraction interlayer. However, as in this case we need to pass worker sets between actions happening at different times PRINGL needs to store the intermediate results (worker sets) that actions scheduled for a future moment will collect when executed (memoization). In case more than one action is scheduled for execution at the same time, the order is unspecified.

Example

The notion of affected workers is important for incentivizing, because the choice of whether or not to perform a subsequent rewarding action may depend on whether previous actions were successfully applied. Consider a company that wants to reward workers either with free days or with a monetary reward. The choice is left to the worker. Free days are offered first. Only workers that refuse the free days will be awarded monetary rewards.

We define a new composite rewarding action `BonusOrDays` (Figure 8.23), which, for the sake of demonstration, assumes the existence of a `RewardAtEndProject` action to award monetary bonuses, as well as a newly defined action `FreeDays` to award free days to the workers.

The output of `a:FreeDays` is the set of workers who accepted the three free days offered. However, due to a complement edge (↛) connecting a and b, the output set of a is subtracted from the original input set. Therefore, the input of `b:RewardAtEndProject` are only those workers who declined to accept free days as reward, and want to be evaluated at the end of project and paid a bonus according to their performance.

Incentive Mechanism ☐

`IncentiveMechanism` is the main structural and functional incentive element. It uses the previously defined complex types to select, evaluate and reward workers of the socio-technical platform. As a self-sufficient and independent unit, it does not have any inputs or outputs. It can be stored and reused through instantiations with different runtime parameters. A complete *incentive scheme* can be specified by putting together multiple incentive mechanisms, prioritizing them, and turning them on/off when needed. Like other complex types, `IncentiveMechanism` also has dedicated GUI elements for definition and instantiation (Fig. 8.24), as well as a shorthand notation used in this book – ⟦IM⟧. Table 11 defines the functionality of ⟦IM⟧'s

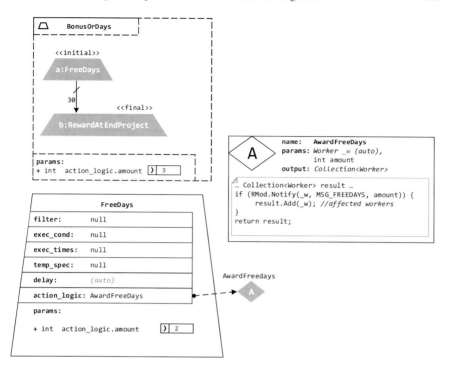

Fig. 8.23: A `CompositeRewardingAction` letting the workers choose one of the rewards.

fields. We show examples of the usage of IMs and other incentive elements in the following section.

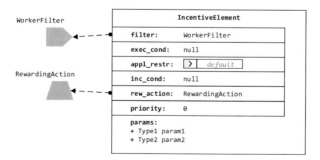

Fig. 8.24: An example `IncentiveMechanism` definition

Field	Description
exec_cond	An optional ⟨P⟩ element used as execution condition for the entire mechanism. Used to check global and time constraints. The condition is commonly used to prevent unwanted multiple executions of the same mechanism. Defaults to true if omitted
appl_restr	Specifies how often a mechanism can be executed in a given interval. The runtime environment then alters the exec_cond accordingly, transparently to the user. This field can be used to turn mechanisms on or off to obtain different incentive scheme configurations
filter	An optional ⟨F⟩ specifying the default target Workers for the ⟨A⟩ specified in field rew_action. If not provided, it defaults to the collection of all the workers in the system. The filter is used to evaluate workers' past or current performance
inc_cond	An optional ⟨P⟩ used to interpret the workers returned by the filter and decide whether to proceed with the rewarding. This condition is meant to be used when the evaluated and targeted worker groups are not the same. In that case, we need to decide whether the results of the evaluation performed through the filter should cause the invocation of the action(s). Returns true if omitted
rew_action	A mandatory ⟨A⟩ assigning the reward or penalty
priority	An optional int indicating the priority of the mechanism's execution. Zero by default

Table 11: Description of IncentiveMechanism fields.

Incentive Schemes (Incentive "Programs")

The *incentive strategy* is the whole of the business logic needed for managing incentives in an organization. The strategy in PRINGL is built bottom-up, by first defining small, reusable chunks of business logic as different complex types. When compiled, the new types are stored in the *incentive library* (Fig. 8.15) with fully qualified names in hierarchical namespaces. From the library they can be instantiated with different parameters and reused in definitions of other complex types, including incentive mechanisms. The mechanisms are combined to obtain an *incentive scheme* (Fig 8.25, ⑤) – a set of high-level rules representing the incentive strategy. An incentive scheme is the equivalent of a visual DSL program. Like any program, it can be run with different parameters and used on different systems with similar characteristics.

Figure 8.25 shows how incentive strategies are constructed. First, missing or specific business logic fragments are defined and compiled into appropriate IncentiveLogic elements (Figure 8.25, ①). In the following steps, after being visually declared, these and other existing library elements can be instantiated for use in definitions of SimpleWorkerFilters and SimpleRewardingActions (Fig 8.25, ②). Similarly, filter and rewarding action type definitions are further used for defining new composite filters and actions (③) and IncentiveMechanisms (④).

Moving up from step ① towards ⑤ the need to know PRINC/PRINGL internals decreases and reduces to understanding the meaning of exposed runtime parameters on a purely visual dashboard. Steps ① - ④ can be skipped altogether if the necessary type definitions are already available from the library. The goal of PRINGL is exactly to promote the reuse of well-defined and common business logic related to incentive management.

Using the graphical elements, the user specifies the necessary runtime parameters for different instances he uses. The GUI environment collects the parameters from all the constituent sub-components and propagates them upwards, possibly until the top-most component's graphical form. The user sets through the GUI whether to propagate a parameter (+/− symbols, Fig 8.25), and therefore delegate the responsibility for filling it out to an upper level, or provide a value at the current level and hide it from upper layers. If a propagated parameter is supplied with different values on different levels, then the rule is that the topmost value overrides all the others. For example, if a parameter is propagated from the ◁▷ level (①) to the incentive scheme level (⑤), then the value defined at level ⑤ is used.

This is possible to do for all the elements that are used to perform predetermined roles (e.g., a *rewarding action* of an IM, or the *auxiliary logic* of an F▷). In this case the runtime environment itself creates the instances and can therefore pass the parameters from the GUI. When the environment does not control the creation of instances (e.g., when IncentiveLogic elements are declared for arbitrary use from code) the programmer must set them in the provided code directly.

8.4.3 Execution Model

The execution of a PRINGL program (incentive scheme) is performed in cycles, as follows.

All ⟦IM⟧s are triggered for execution whenever a *triggering signal* from the abstraction interlayer is received. It is the responsibility of the Designer to ensure through priorities and execution conditions that a specific order of execution of ⟦IM⟧s is achieved. The order of execution of ⟦IM⟧s with the same priority is not predetermined. Execution conditions of the ⟦IM⟧s with higher priorities are evaluated first. Only after the higher-priority ⟦IM⟧s have executed are the conditions of lower-priority ones evaluated. This allows the higher-priority mechanisms to preemptively control the execution of lower-priority ones by changing condition variables through side effects. The execution time of any single ⟦IM⟧ is limited by design to a maximum time T_{IM}^{max}, which is the time needed to pass the message to the underlying socio-technical platform. Therefore, a single execution cycle of an incentive scheme of n mechanisms can last at most $T_{sc}^{max} = n \times T_{IM}^{max}$. It is necessary that $T_{sc}^{max} < T_{tick}$, where T_{tick} is the basic time of the abstraction interlayer (*tick* in case of PRINC). The execution of an ⟦IM⟧ begins by evaluating exec_cond. If true, the associated filter is passed the collection of all the workers in the system and invoked. The resulting workers are then passed to the incentive_cond to decide whether the execution should proceed with rewarding. If it returns true, rew_action is invoked. If the action does not override its filter field PRINGL passes the collection of workers returned by the ⟦IM⟧'s filter field.

An ⟦F⟧ executes by checking for each worker from the input collection whether he fulfills the provided predicate. This is done for each PoiT returned by time_restr (⟨↻⟩). The results are then interpreted in accordance with the provided temp_spec. For example, if the specifier is Once() then it suffices that the worker fulfilled the predicate in at least one of the PoiTs in order to be placed in the resulting collection. In case of composite filters the constituent sub-filters are executed in the defined order. The initial sub-filter (marked <<initial>>) receives the initial collection of workers from the environment, which is then passed on to subsequent filters. The resulting collection of workers from the <<final>> sub-filter is returned as the overall result. The <<initial>> filter is given different default inputs by the PRINGL environment depending on where the composite filter is instantiated. The anonymous :Passthru sub-filter instances are special PRINGL sub-filter types passing the union of workers from all input edges on all output edges without performing any filtering.

A simple ⟦A⟧ is executed if the exec_cond (⟨P⟩) returns true. In this case, the execution PoiTs for the action are obtained from exec_times (⟨↻⟩) and then interpreted in accordance with the temp_spec. Once the times are determined, the environment schedules the action in the abstraction interlayer (in our case PRINC's *Timeline*) and provides the actual code that performs the action from the action_logic (⟨A⟩). However, during the entire runtime PRINGL keeps track of the scheduled action, in order to honor temporal specifications and to detect rescheduling due to Interval redefinitions. The workers to whom the action applies are taken from the associated

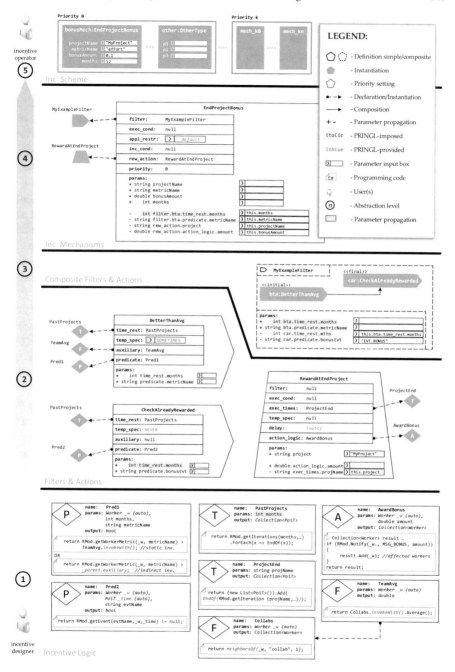

Fig. 8.25: Incentive scheme from Example 8.4.4.3, illustrating the decrease in complexity going from modeling of (low-level) incentive elements by incentive designers to adjusting existing incentive schemes by incentive operators

filter. As explained, if the local filter is omitted, PRINGL defaults to the workers from the parent 𝕀𝕄's filter.

The execution of a composite action starts by first breaking it into linear execution paths containing constituent simple actions. For each execution path PRINGL takes into account specified delays and adjusts the ⟨T⟩ elements in constituent actions to account for provided delays, which are then (re-)scheduled with the abstraction interlayer. However, as in this case we need to pass worker sets between actions happening at different times PRINGL stores the intermediate results (worker sets) that actions scheduled for a future moment will collect when executed (memoization). In case more than one action is scheduled for execution at the same time, the order is unspecified.

Executing incentive logic elements ⟨L⟩ results in invoking the instance similarly to a conventional function. The environment passes both the auto parameters and any user-defined ones. If user-defined parameters are omitted when an ⟨L⟩ is invoked from the code by indirect invocation the parameters are obtained from the visually exposed parameter fields. However, when supplied, the arguments provided in the code override those provided in the fields. If the parameter value cannot be resolved in either way, the invocation fails.

Parameters are collected and propagated automatically from instances created to fulfill complex type field roles. In that case the runtime environment controls the instantiation and therefore knows to which instances to pass the parameters from the GUI. When the environment does not control the creation of instances (e.g., when IncentiveLogic elements are declared for arbitrary use from code) the programmer must set them in the provided code directly.

Overall, PRINGL's execution is "best effort". This means that PRINGL expects the interlayer to pass to the underlying socio-technical system the rewarding actions to be taken, but will not expect them necessarily to be observed. Acknowledgments are used to keep track of successfully applied rewarding actions. If any error is encountered during the execution, the currently invoking incentive mechanism fails gracefully, but the execution of other mechanisms continues. The incentive scheme's execution needs to be stopped explicitly.

8.4.4 Evaluation

A domain-specific language (DSL) can be evaluated both quantitatively and qualitatively. *Quantitative analysis* of the language is usually performed once the language is considered mature [113], since this type of evaluation includes measuring characteristics such as productivity and subjective satisfaction, which require an established community of regular users [160]. We therefore evaluate PRINGL qualitatively, which, in general, can include comparative case studies, analysis of language characteristics and monitoring/interviewing users. Analysis of language characteristics was chosen as the preferred method in our case, since it was possible to perform it on the basis of the findings gathered through analysis of numerous existing incentive models and presented in Section 8.2.4. Due to difficulties in engaging a relevant number of domain experts willing to take part in monitoring we were unable to perform this type of user-based evaluation at this point. Comparative analysis was not applicable in this case, due to nonexistence of similar languages. The evaluation is performed with respect to the language requirements elicited in Section 8.4.1.1. We constructed an example suite covering realistic incentive elements identified in Chapter 8.2.4. By implementing and analyzing different incentive use cases from the suite we showcase the usage of PRINGL and argue for the coverage of the requirements. Concretely, the requirements are evaluated as follows:

- The diversity of examples in the suite and the fact that they were obtained from the broad survey of realistic incentive practices testify to PRINGL's *groundedness* and *expressiveness*.
- Through elaborate discussion of particular implementation details of different suite examples we demonstrate PRINGL's *reusability* and *portability*.
- While lacking the necessary conditions and metrics to conclusively show the *usability* of the language, the implemented set of examples allows us to conclusively argue for certain aspects of usability, such as "*usefulness*" and "*portability*" (as defined in [160]).

Table 12)[17] shows the coverage of the chosen examples with respect to introduced incentive categories and their constituent parts. Some examples are presented partially to illustrate/highlight the claimed capabilities that the particular example is supposed to cover.

8.4.4.1 Example – Employee Referral

A company introduces an employee referral process[18] *in which an existing employee can recommend new candidates and get rewarded if the new employee spends a year in the company having exhibited satisfactory performance.*

Solution: In order to pay the referral bonuses (deferred compensation) the company needs to: a) identify the newly employed workers; and b) assess their subsequent performance. Let us assume that the company already has the business logic for assessing the workers implemented, and that this logic is available as the library filter GoodWorkers. In this case, we need to define one additional simple filter NewlyEmployed, and combine it with the existing GoodWorkers filter. In Figure 8.26 we show how the new composite ReferralFilter is constructed. The 𝔽 instance n:NewlyEmployed makes use of: a) ⟨𝕋⟩ PastMonths returning PoiTs representing end-of-month time points for the given number of months (12 in this particular case); and b) predicate ⟨ℙ⟩ Pred2 checking whether the employee got hired

---

[17] Note that the Indirect and Subjective evaluation methods have been omitted from Table 12. The former, because it implies use of sophisticated evaluation algorithms, but implementation-wise would not differ from the Quantitative evaluation. The latter, because is not easy to uniformly model in software, as it implies subjective human opinions that are unknown at design time.

[18] http://en.wikipedia.org/wiki/Employee_referral

	Ex.8.4.4.1	Ex.8.4.4.2	Ex.8.4.4.3	Ex.8.4.4.4	Ex.8.4.4.5
Incentive Category					
PPP			✓		
Quota/Discretionary			✓		
Deferred Compensation	✓				
Relative Evaluation			✓		
Promotion					✓
Team-Based Compensation		✓			
Psychological				✓	✓
Rewarding Action					
Quantitative		✓	✓		
Structural					✓
Psychological				✓	✓
Evaluation Method					
Quantitative			✓	✓	✓
Peer Voting		✓			

Table 12: Coverage of incentive categories, rewarding actions and evaluation methods by the provided examples

12 months ago. Pred2's general functionality is to check whether the abstraction interlayer (RMod) registered an event of the given name at the specified time.

Discussion: The shown implementation fragment illustrates how easy it is to expand on top of the existing functionality. Under the assumption that there exists a metric for assessing the workers' performance, and that it can be queried for past values (cf. PRINC's Timeline), introducing the employee referral mechanism is a matter of adding a handful of new incentive elements.

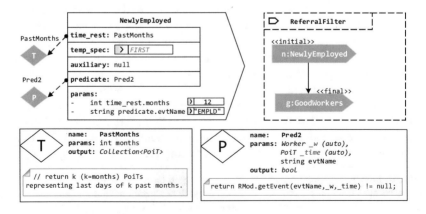

Fig. 8.26: A CompositeWorkerFilter for referral bonuses

8.4.4.2 Example – Peer Voting

Equally reward each team member if both of the following conditions hold: a) each team member's current effort metric is over a specified threshold; and b) the average vote of the team manager, obtained through anonymous voting of his subordinates, is higher than 0.5 [0–1].

Solution: As shown in Fig. 8.27 we compose an incentive scheme consisting of two [IM]s – i1:PeerAssessIM, in charge of peer voting; and i2:RewardTeamIM in charge of performing team-based compensation. [IM] i1 will execute first due to its higher priority, and set the global variable done, through which the execution of i2 can be controlled (⟨P⟩ PeerVoteDone). [IM] PeerAssessIM uses the [F] TeamMembers to exclude the manager from the rest of the team members. The filter TeamMembers is a composite filter composed of two subfilters [F] GetManager and [F] GetTeam, borrowed from Ex.8.4.4.5, Fig. 8.30. The resulting workers are passed to [A] DoPeerVote, which performs the actual functionality of peer voting. The referenced rewarding action is simple; it just passes to ⟨A⟩ PeerVote the workers who need to participate. The ⟨A⟩ PeerVote is performed by dispatching messages to the workers and receiving and aggregating their feedback through the abstraction interlayer. Once the peer voting has been performed, the manager's assessment is stored in _global.mark, and the flag _global.done is set to allow execution of [IM] i2. Once set to execute, the [IM] i2 first reads all the team members via [F] GetTeam. Whether they ultimately receive the reward depends on the evaluation of the inc_cond field. The field contains a conjunction of two indirectly invoked ⟨P⟩ elements (Sec. 8.4.2.4). The condition expresses the two constraints from the incentive formulated in natural language. If it resolves to "true," the [A] DoRewardTeam applies a predefined monetary reward, sharing it equally among all team members (via ⟨A⟩ RewardTeam).

Discussion: The key question here is how to support incentives requiring direct human feedback, such as peer voting. Such interactions require support from the abstraction interlayer. To support this functionality, the abstraction interlayer can either rely on the functionality offered by the underlying platform, or provide this functionality independently to safeguard voting privacy and incite expression of honest opinions. In earlier chapters we presented SMARTCOM – a framework for virtualization and communication with human agents. In this example we model the latter variant in PRINGL, assuming the use of PRINC with SMARTCOM for interaction with workers.

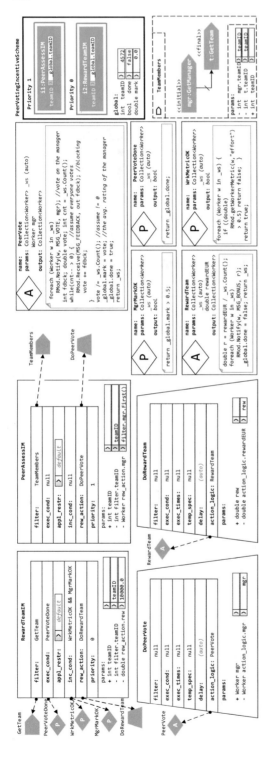

Fig. 8.27: An incentive scheme example combining peer voting and team-based compensation

8.4.4.3 Example – Bonus

Award a 10% bonus to each worker W who sometimes in the past 12 months had a higher value of metric "effort" than the average of workers related to W via a relationship of type "collab", and who was not rewarded in the meantime.

Solution: Figure 8.25 shows the bottom-up implementation of this incentive (①-⑤). First, at level ① we define novel or context-specific business logic fragments as IncentiveLogic ◁Ɫ▷ elements. This level relies on the abstraction interlayer to read the updated worker metrics, obtain data about recorded events and send system messages. At ② we define new F▷ and /A\ types. Similarly, F▷ and /A\ definitions are further used for defining new composite filters and actions (③) and IncentiveMechanisms (④). By setting the parameter fields the designer specifies the necessary runtime parameters for different instances. Apart from constants, a field can contain references to other fields "visible" from that element. The environment collects the field values (parameters) from all the constituent sub-components and propagates them upwards, possibly until the topmost component's GUI form. Through the +/– symbols the designer controls whether to propagate a parameter and, thus, delegate the responsibility for filling it out to the higher level, or provide a value at the current level and hide it from higher levels. Parameter propagation is one of PRINGL's usability features. In Fig. 8.25 we show an example of parameter propagation (marked in orange). Element ◁Ɫ▷ PastProjects (①) exposes the parameter months. The same parameter is then re-exposed by F▷ BetterThanAvg (②), which uses PastProjects as its time restriction. The parameter is further propagated up through F▷ MyExampleFilter until it finally gets assigned the value in ⓘⓂ EndProjectBonus (④).

Discussion: This incentive mechanism was chosen to highlight a number of important concepts. Every underlined term in the natural language formulation of this incentive mechanism is a specific value of a different parameter that can be changed at will. In PRINGL terms, this means that the incentive operator can easily switch between different (library) incentive elements of the same type/signature and tweak the parameters to obtain different incentive mechanism instances. In this way, incentive designers or operators can adapt generic mechanisms to fit their needs. If we analyze the generic version of this incentive mechanism, we can see that it embodies the principles of pay-per-performance incentives based on the value of a quantifiable metric, but coupled with an additional condition that is evaluated relative to co-workers. In addition, the mechanism contains two temporal clauses ("in past 12 months" and "in the meantime"), making it also a representative of the quota system type of incentive.

The example also demonstrates reusability – the ◁Ɫ▷ PastProjects is reused twice in two different F▷s. Also, steps ①–④ can be skipped altogether if the necessary type definitions are already available from the incentive library. As we can see, at levels ②–⑤ only visual programming is required. This means that there is no need to know any interlayer internals, apart from understanding the meaning of propagated parameters. So, if different platforms offer standardized implementations of the commonly used incentive logic, the incentive elements become completely portable.

8.4.4.4 Example – Rankings

Let us assume that the imaginary platform from Example 8.4.4.3 wants to extend the existing incentive scheme with an additional incentive mechanism in an (admittedly over-simplified) attempt to raise the competitiveness of underperforming workers:

Show the list of the rewarded employees and their performance (rankings) to those workers that did not get the reward through application of ⬚IM⬚ EndProjectBonus *in Ex. 8.4.4.3 (Fig. 8.25).*

Solution: Figure 8.28 shows the additional elements needed to support the new mechanism. The composite ⬚F▷ NonRewardedOnes reuses the existing ⬚F▷ MyExampleFilter from Ex. 8.4.4.3 as initial subfilter, and returns the set complement, i.e., the non-rewarded workers to whom the rankings need to be shown. In order to display the rankings, we copy and paste the existing ⬚A\ RewardAtEndProject from Ex. 8.4.4.3 and change only the value of the field action_logic to point to the the newly defined ◁A▷ ShowRankings, also shown in Fig. 8.28. Let us name the newly obtained ⬚A\ RankingsAtEndProject. In the same fashion, we copy and paste the existing ⬚IM⬚ EndProjectBonus from Ex. 8.4.4.3, make its filter and rew_action fields point to the newly defined ⬚F▷ NonRewardedOnes and ⬚A\ RankingsAtEndProject, respectively. The obtained ⬚IM⬚ performs the required functionality.

Discussion: This example shows a common, realistic scenario, where additional incentive mechanisms need to be added to complement the existing ones. In this case, the added mechanism acts on the underpeforming workers psychologically by showing them how they fare in comparison to the rewarded workers. Such mechanisms can be used to motivate better-performing underperformers, while having a de-motivating effect on the worst performing ones. As we have shown, such a mechanism can be easily and quickly constructed in PRINGL with minimal effort.

8.4.4.5 Example – Rotating Presidency

Teams of crowd workers perform work in iterations. In each iteration one of the workers acts as the manager of the whole team. This scheme motivates the best workers competitively by offering them a more prestigious position in the hierarchy. However, in order to keep team connectedness in the longer run and foster equality and fresh leadership ideas, a single person is prevented from staying too long in the managerial position. Therefore, in the upcoming iteration the team becomes managed by the currently best-performing team member, unless that team member was already presiding over the team in the past k iterations.[19]

Solution: For demonstration purposes, we are going to fully model the type definitions necessary for implementing the rotating presidency incentive scheme. However, in practice it is reasonable to expect that a significant number of commonly

---

[19] An iteration can represent a project phase, a workflow activity or a time period.

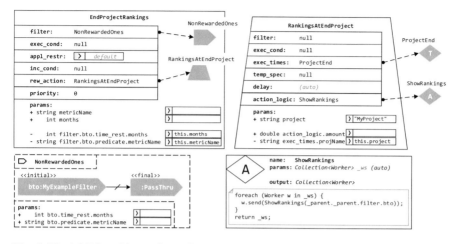

Fig. 8.28: Additional incentives elements needed to augment the incentive scheme from Example 8.4.4.3 (Fig. 8.25) in order to display motivational rankings to the non-rewarded workers from Example 8.4.4.3

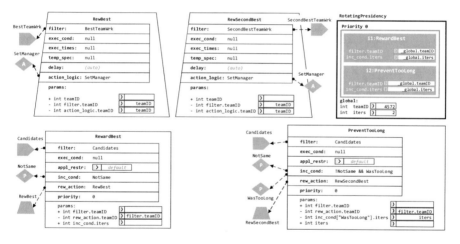

Fig. 8.29: Modeling the rotating presidency incentive scheme in PRINGL. Segment showing the incentive scheme (top right), rewarding actions (top center and left), and incentive mechanisms (bottom)

used type definitions would be available from a library, cutting down the incentive modeling time.

Contrary to Example 8.4.4.3, this time we adopt a top-down approach in modeling. In order to express the high-level functionality of the rotating presidency scheme the Designer uses PRINGL's visual syntax to define an incentive scheme named RotatingPresidency (Fig 8.29, top right) containing (referencing) two IM instances – i1 and i2, with the same priority (0). The RotatingPresidency scheme definition also contains a set of global parameters that are used for configuring the execution of the scheme: teamID uniquely defines the team to which we want to apply the scheme, while iters specifies the maximum number of consecutive iterations a team member is allowed to spend as a manager. By choosing different parameter values an incentive operator (Operator) can later adjust the scheme for use in an array of similar situations in different organizations.

The two incentive mechanisms that the scheme references; i1 and i2, are instances of the ⟦IM⟧ types RewardBest and PreventTooLong, respectively (Fig 8.29, bottom). The ⟦IM⟧ RewardBest installs the best worker as the new manager if (s)he is not the manager already. The ⟦IM⟧ PreventTooLong will replace the current manager if the worker stayed too long in the position, even if the manager was again the best performing team member. "Installing" or "replacing" a manager is actually performed by re-chaining of management relations in the structural model of the team by applying appropriate graph transformations [82] through the abstraction interlayer.

When the incentive condition (inc_cond field) of ⟦IM⟧ PreventTooLong evaluates to true, this means that the current manager has occupied the position for too long, and should now be replaced by the second-best worker. PRINGL does this by invoking the specified ⟨A⟩ RewSecondBest and passing it the collection of workers returned by the ⟨F⟩ Candidates. The ⟨F⟩ Candidates returns potential candidates for the manager position – the best-performing Worker and the current manager. The same filter is referenced from both ⟦IM⟧s. However, the ⟦IM⟧ PreventTooLong invokes ⟨F⟩ Candidates through a complex incentive condition field, referring to two ⟨P⟩ elements, which both need to be visually declared. PRINGL allows this as a shorthand notation instead of forcing the user to create a container ⟨P⟩ element to perform the same logical function. In this case, the exposed parameters cannot be simply referenced by using the field name, but rather the parameters are accessed through an associative array (C# Dictionary) bearing the same name as the field, while the names of the used ⟨P⟩ elements serve as key names. For example, to access the ⟨P⟩ WasTooLong's parameter iters from ⟦IM⟧ PreventTooLong where ⟨P⟩ WasTooLong is used in the inc_cond field, we must write: inc_cond["WasTooLong"].iters As it can be visually tiring to read the lengthy fully qualified names of propagated parameters, we often stop propagating such parameters and propagate a new, local one with the same name, whose value we then copy to the long-named parameter (e.g., just iters instead of inc_cond["WasTooLong"].iters).

Both ⟦IM⟧s get executed always as the nullified exec_cond fields default to true. However, ⟦IM⟧ PreventTooLong's incentive condition (inc_cond field) contains: !NotSame && WasTooLong. It ensures that the ⟨A⟩ RewSecondBest of ⟦IM⟧ PreventTooLong will never get executed at the same time as the ⟨A⟩ RewBest of the ⟦IM⟧ RewardBest.

Two rewarding actions are instantiated and invoked from the ⟦IM⟧s. The ⟨A⟩ RewBest monitors the "effort" metric and rewards the best worker in the current iteration. The ⟨A⟩ RewSecondBest replaces the current team manager with the second-best performing worker when needed. The ⟦IM⟧ inc_cond fields make sure that the two actions do not get executed in the same iteration. The fact that a rewarding action

instantiates its own `filter` means that it discards the workers passed to it by the
PRINGL environment from the encompassing ⟦IM⟧'s `filter` field and rewards those
returned by the local filter.

In both actions most fields are nullified, meaning that the PRINGL execution
environment will assume the default field value. This means that the `action_logic`
⟨Ⓐ⟩ SetManager will be unconditionally scheduled for execution.

We now show how the previously referenced filters are defined. We will first
describe the definitions of the three simple filters (Fig 8.30, right) and then use them
to visually assemble the definitions for another four composite filters (Fig 8.30, left).

- GetTeam: Returns all the workers belonging to the team with the specified
  `teamID`. The filtering is performed by running each of the workers from the
  input set against the `predicate` ⟨Ⓟ⟩ IsTeamMember and including the worker
  in the output if he fulfills the predicate.

- GetBest: Returns the worker who has achieved the highest value of the 'effort'
  metric by invoking the ⟨Ⓕ⟩ GetWrkBestMetric and then just formally matching
  him with the `IsBest` predicate. In this example we use the 'effort' metric [144],
  but any other compatible performance metric could have been used and exposed
  as a global parameter. This filter does not care to which team the evaluated
  worker belongs – if used independently, it evaluates all the workers in the system.
  This is why we always use it in composite filters, where we initially restrict its
  input set with another filter.

  In our example this filter encapsulates and hides the metric it uses for evaluating
  the workers. In principle, it makes sense to propagate the metric name upwards
  and thus make it user-settable, consequently making the whole scheme more
  general. However, for readability purposes we decided not to propagate this
  parameter in this example.

- GetManager: Invokes an ⟨Ⓕ⟩ GetMgrByRelations that performs a graph query
  [82] on the team model through the abstraction interlayer to determine the
  manager within the provided input set of workers.

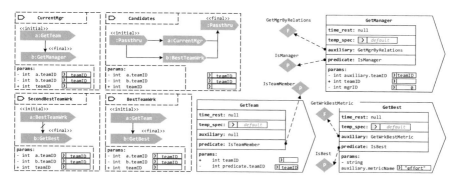

Fig. 8.30: Modeling the rotating presidency example: Segments show simple filters
(right) and composite ones (left)

Composite filter type definitions are constructed visually. The following composite filters are defined:

- CurrentMgr: Returns the current manager of the team. The $\mathbb{F}\!\triangleright$ a:GetTeam returns all the workers belonging to the team with the teamID, while the $\mathbb{F}\!\triangleright$ b:GetManager uses managerial relationships to determine the manager among those workers.[20]
- BestTeamWrk: Returns the best individual from a previously identified collection of team members. The $\mathbb{F}\!\triangleright$ b:GetBest determines what "best worker" means in this case.
- SecondBestTeamWrk: As the name suggests, returns the second-best worker in the team. The subfilter a returns the best worker of the team and passes it forward to the subfilter b via a negated edge ($\nrightarrow$). This means that b now receives as input: *input(a)* \ a, i.e., in this particular case the collection of all workers belonging to the team minus the best worker. Subfilter b returns the best worker from this collection, and thus effectively the second-best worker of the team.
- Candidates: This filter simply uses the previously defined filters CurrentMgr and BestTeamWrk and returns the set union of their results.

---

[20] While managerial relations in principle need not be stored as a graph, and can thus be identified much more easily, we still use the graph managerial relations as an easily understandable example of how any graph-encoded structural property can be used in incentive management.

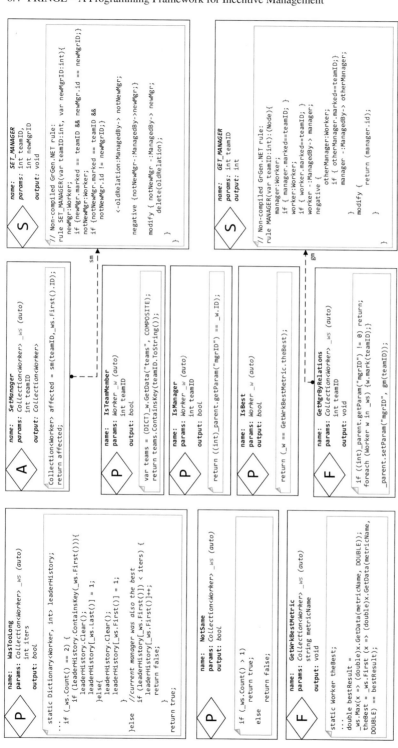

Fig. 8.31: Modeling the rotating presidency example. Segment showing the incentive logic elements

The incentive logic elements, shown in Figure 8.31, contain the low-level business logic and code[21] that communicates with the abstraction interlayer. The Designer takes implements the incentive logic elements as small code snippets with intuitive and reusable functionality. A short description of the functionality of the employed elements is provided in Table 13.

Element	Symbol	Description
IsTeamMember	⟨P⟩	Determines whether a given worker belongs to a given team
IsManager	⟨P⟩	Checks whether the currently evaluated worker has the ID previously determined to belong to the team manager by ⟨F⟩ GetMgrByRelations
IsBest	⟨P⟩	Checks whether the currently evaluated worker is the same as the one identified by the GetWrkBestMetric
NotSame	⟨P⟩	Determines whether the input contains two manager candidates
WasTooLong	⟨P⟩	Keeps track of how many times a worker was in the manager position, and returns true if the worker is not supposed to become manager in the upcoming iteration
GetWrkBestMetric	⟨F⟩	Reads the value of the effort metric for each of the passed workers in _ws and updates the best worker
GetMgrByRelations	⟨F⟩	Invokes the read-only structural query ⟨S⟩ GET_MANAGER
SetManager	⟨A⟩	Invokes the modifying structural query ⟨S⟩ SET_MANAGER
GET_MANAGER	⟨S⟩	Contains a compiled non-modifying GrGen.NET graph query, here expressed in the GrGen rule language. The rule only considers the nodes marked by the teamID tag (see GetMgrByRelations). The rule matches and returns a node that other nodes point to via ManagedBy-typed relations, but itself is not managed by another team member
SET_MANAGER	⟨S⟩	Contains a compiled modifying GrGen.NET graph query matching the old and the new manager, and re-chaining the ManagedBy relations to point to the new manager node

Table 13: Incentive logic elements used in the rotating presidency example

Discussion: This example combines the promotion and psychological incentives. Promotion is performed through a structural rewarding action, and is designed to foster competitiveness and self-esteem. At the same time, team spirit and a good working environment are being promoted by limiting the number of consecutive terms, thus giving a chance to other team members. This example shows a fully implemented and executable incentive scheme. Although the model may seem complex at the first glance, it is worth noting that the type definitions of the two actions (Fig 8.29, top) are almost identical, differing only in the filter they use – with the first using the ⟨F⟩ BestTeamWrk and the second the ⟨F⟩ SecondBestTeamWrk. This

---

[21] Here we use C# in all but ⟨S⟩ elements, which are shown in the original GrGen.NET rule language: http://www.info.uni-karlsruhe.de/software/grgen/

means that once the Designer has modeled one of them, the other one can be created by copying and pasting and referencing a different filter. Similarly, if at a later time the underlying platform decided to use a different ⟨A⟩ to reward the best workers (e.g., to pay out money instead of rotating team managers) the Designer would only need to partially adapt the scheme by referencing a different ⟨A⟩ from the ⟨A⟩'s action_logic fields. Such adaptations can also be performed by incentive operators with minimal understanding of the underlying code.

Filters such as GetTeam, GetBest and GetManager perform very common incentive functionality. In practice, this means that such components may be readily available as library elements. Of course, if we need to use a company-specific flavor, we can easily replace the default one with a proprietary element. For example, an F⟩ GetManager may be available with a default auxiliary field ⟨F⟩ that looks for a manager in the team model by inspecting the node tags for a given manager tag. In that case, to adapt such a filter for our rotating presidency example the Designer needs to replace the default, tag-based ⟨F⟩ with a structural one, such as GetMgrByRelations.

8.4.5 Implementation

Figure 8.15 (Sec. 8.4.1) shows an overview of the implemented components. PRINGL's language metamodel was implemented in Microsoft's Modeling SDK for Visual Studio 2013 (MSDK). Source code, screenshots and additional info is available for download[22] MSDK allows visual DSLs to be defined and translated into an arbitrary textual representation. Using MSDK we generated a Visual Studio plug-in providing a complete IDE for developing PRINGL projects. In it, an incentive designer can create a dedicated Visual Studio PRINGL project and implement/model real-world strategies using the visuo-textual elements presented in this book (Figure 8.32). The graphical elements provided in the implemented Visual Studio PRINGL environment, although not as visually appealing as those presented here, functionally and structurally match them fully. PRINGL models are stored in .pringl files that get automatically transformed into the corresponding C# (.cs) equivalents. The generated code can then be used in the rest of the project as regular C# code or compiled in .NET assemblies (e.g., libraries or executables).

As a proof of concept, demonstrating the feasibility of implementation of the introduced programming and execution model, we implemented the "rotating presidency" example (Ex. 8.4.4.5) from Section 8.4.4.5. Figure 8.32 shows a screenshot of implemented rotating presidency example using the VS PRINGL IDE as well as the intended use of generated code artifacts. The implemented incentive elements correspond to the individual element descriptions presented in example Ex. 8.4.4.5. The entire scheme was modeled using the generated PRINGL tools, demonstrating the feasibility of the proposed architectural design. The C# code obtained from the

---

[22] http://dsg.tuwien.ac.at/research/viecom/PRINGL/

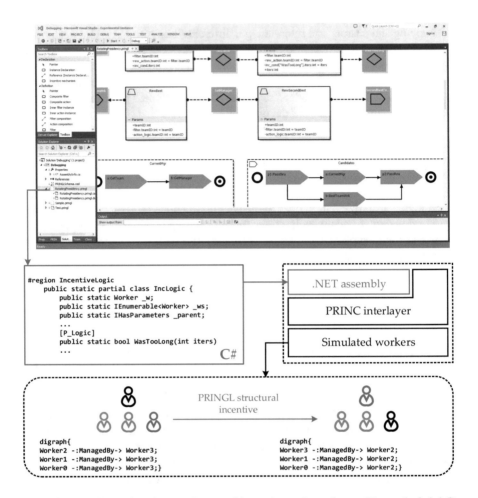

Fig. 8.32: Implementing the rotating presidency incentive scheme (Example 8.4.4.5) using the PRINGL Visual Studio environment. Generated C# code performs calls to PRINC APIs, which ultimately perform structural changes on the worker graph (part of the RMod)

implemented model can be used to produce a custom-made incentive management application using PRINC as the abstraction interlayer.

The implemented example supports an arbitrary number and structure of Workers (represented as graph nodes) and their "effort" metrics. Worker nodes are interconnected with arbitrarily typed graph edges representing different relations. Our PRINGL-encoded incentive scheme will only consider the workers belonging to the team denoted by the teamID identifier, and only the managerial relations represented by ManagedBy-typed edges. Events notify PRINC when iterations end and "effort" metrics change. The code generated from the implemented example monitors these events and executes the incentive mechanisms that make sure the best-performing worker is installed as the manager, but for not more than two consecutive iterations, subject to being replaced by the runner-up in such a situation.

The evaluation of PRINGL's performance was not of interest to us at this phase. Furthermore, in the absence of any related domain-specific languages or modeling approaches, no comparative analysis was possible.

8.4.6 Summary

Throughout this chapter we have argued for the necessity of composition and frequent adjustments of incentive elements and mechanisms. In this respect, the roles of Incentive Designer and Incentive Operator may be seen as critical to the success of any future Smart City value-added application relying on citizen-provided effort through the city's socio-technical platform. For companies/organizations that will operate the value-added applications, one of the major concerns will be to control the effectiveness of the incentives and the costs associated with applying them.

In this chapter we introduced an incentive management framework to support incentive designers and operators in designing, adapting and applying incentive schemes, allowing quick adaptations to negative behavioral trends. The design has been evaluated for the principal usability requirements. The associated domain-specific language PRINGL was also introduced. Its visual syntax, the programming and the execution model were described and discussed using a number of examples. The language allows very generic and portable encoding of composite incentive strategies that are independent of the number of managed peers, lending itself well to application in Smart City domains.

Part IV
Towards the Smart City of the Future

Preface

Throughout this book we have presented our vision of future developments of the Smart City concept based on novel advances in the areas of Cloud Computing, Internet Of Things and Social Computing. The new Cyber-Human Smart City offers the possibility to blend these technologies in a single application environment and manage them in a coordinated manner through a unified platform. This synergy allows more efficient ("smarter") management of the city's infrastructural resources and human capital, including citizens' privately owned property (sensors, actuators, computing/storage units, and any IoT-enabled object offered for shared use) and their own physical and cognitive abilities. Even more importantly, this synergy offers novel possibilities to capitalize on personal capabilities and belongings, to learn about and take an active part in the city's decision-making and to engage with fellow citizens in various collective activities in both virtual and physical domains. In this final chapter we explore the benefits of horizontal integration of the described technologies and present a research road map identifying the next important steps in the development of Cyber-Human Smart Cities.

Chapter 9
A Road Map to the Cyber-Human Smart City

In Chapter 1 we have described and structured these benefits in an *Architecture of Values*. We have discussed how the proposed Architecture of Values can serve as a conceptual framework for a value generation process in Smart Cities of the future. We have also presented a conceptual architecture of a Smart City platform that acts as the technological basis for the embodiment of these values and the value generation process. In subsequent chapters (Part II and Part III of this book), we presented a number of the platform's key technologies, based on our previous work in this field, and described the recent advances in these areas that are necessary for supporting the functionality of the platform. In this chapter we discuss how the introduced technical solutions can work together to facilitate the value generation process, and map them to particular values and problems introduced in Chapter 1. Finally, we discuss the next steps in realizing the stated Cyber-Human Smart City vision.

9.1 Going Beyond the Contemporary Smart City – Horizontal Integration as a Value Generator

The contributions described in Chapters 3–8 of this book individually offer solutions to important research/technological challenges in the domains of Internet of Things and Social Computing. However, only when these solutions are considered holistically do their potential benefits become obvious in the context of Smart Cities. In the following we discuss, based on several practical examples, how the introduced solutions can be horizontally integrated to help build our vision of the Cyber-Human Smart City centered around the value generation architecture (Fig. 1.1). The values the horizontal integration generates span all three value domains (infrastructural, societal, business), going beyond contemporary Smart City goals and promising to make the cities of the future smarter and more livable.

© Springer International Publishing AG 2017
S. Dustdar et al., *Smart Cities*,
DOI 10.1007/978-3-319-60030-7_9

9.1.1 Managing and Exploiting Diversity

Probably the biggest practical challenge that all contemporary Smart City visions need to address is the scale and diversity of infrastructural elements that need to be managed. While the scale is more of a technical problem, the diversity is a problem that fundamentally hinders the development of the Smart City vision, since it forces the city management to make hardware and software vendor choices and thus establishes a centrally managed system, suffocating the liberal market and healthy competition. Therefore, instead of citizens freely deciding on the equipment they want to use and connect to the city's grid and infrastructure management platform, it is usually the city or a private company that establishes control centers that manage the IoT infrastructure of a limited number of supported manufacturers and enforce the safety, privacy and compliance policies. In a similar manner, managing humans in existing social-computing platforms is usually scenario/application-centric. This means that different platforms are not interoperable, and that a worker's reputation and experience cannot be transferred as evidence to other platforms, nor can a platform be easily extended to support different types of labor. This is why the majority of today's social-computing platforms are crowdsourcing platforms offering only simple tasks that can be solved individually and with limited expertise. This means that contemporary social-computing platforms are not a serious contender to traditional labor markets and cannot attract people to consider building professional careers on social-computing platforms.

All the solutions presented in Parts II and III of this book were designed to support diversity in an effort to alleviate such problems and allow management of diverse hardware and human resources. Going back to the problem of free choice of equipment manufacturer, our Smart City vision supports incentivizing citizens to come together in collectives (e.g., representing particular buildings, neighborhoods or interest groups) to discuss and decide on the equipment they want to acquire/install. The choices may require different kinds of agreements and compromises, as the potential benefits from the installed equipment will never be equal for all the affected citizens. Based on this, the Smart City platform can offer different incentives to try to get particular citizens "on board" and grow the IoT/infrastructure network. Of course, which incentives would be employed in this case depends on the perceived or calculated benefit that the city expects from the subsequent use of the equipment.

In case the installed equipment is, e.g., a "smart car controller" (a device able to remotely locate the car and monitor emissions; possibly start, stop and restrict movement of the vehicle), the benefit to the city is the ability to better plan and manage the overall city traffic. Since the potential savings in the city budget can be significant compared to the costs of incentives, the city invests in an incentive scheme to stimulate citizens to buy and install such devices. The citizens are free to choose different device manufacturers, since the Smart City platform is able to provision, run and communicate with the different software instances on different devices, due to the standardized API and model of the device.

It is not difficult to visualize the potential infrastructural values this could bring. It would gradually allow a logically centralized management of the complete traffic infrastructure of the city, including the privately owned vehicles. This means that the traffic flow could be optimized and emissions reduced. Citizens could be motivated to take specific routes, use public transport or ride-sharing, offer rides to others, or purchase pre-assigned CO_2 quotas of others. Existing "brute-force" methods taken in extreme cases of air pollution, such as license-plate rationing, are often inefficient.[1] The described scheme, if underpinned by the majority of drivers in a city, could solve this problem, in that it could control the traffic on a finer-grained level.

This brings us to the second point – the success of such a scheme is dependent on both the high number of participating citizens and their willingness to embrace a system where everyone sacrifices a piece of perceived freedom and privacy in order to build a system that caters for the common good, and in the long run brings more benefits to everyone. In this particular case, the citizen occasionally sacrifices the freedom of movement with the personal vehicle (with respect to space, time, and passenger restrictions), but gains in less congested streets and decreased pollution. While in theory this sounds viable, in practice the phenomenon known as the "tragedy of the commons"[2] prevents the usability of the established scheme. This is why our proposed approach relies on incentives, which serve not only to motivate increased participation levels, but also to fight free-riding (dysfunctional behavior), as explained in Chapter 8.

In this particular case, the citizens are being motivated to voluntarily install the smart-car controllers and accept the participation rules by being offered tangible benefits in terms of significantly reduced transportation costs in the long run (discounted public transport, free ride-sharing, reduced registration costs, fuel tax refunds) upon demonstrating active participation in the system. Dysfunctional behavior is easily tracked and can be penalized, but the rules must be designed only to penalize repeated and planned gaming of the system. Occasional breaking of the rules must be embraced and not penalized, to accommodate for unforeseeable human necessities. It is important for the system not to be perceived by the citizens as disruptive and too restrictive. In the long run, however, each participant must be aware that the system is fair and beneficial to every honest participant. Apart from a clear incentive scheme and participation rules, from the technical point of view, the fact that the SmartCity platform is the sole point of control provisioning and managing the software executed on the controllers is important to maintain faith in the fairness of the overall system.

While for the city's management the infrastructural value produced by the described scheme is important, for the normal citizens the scheme produces a clear societal value, in that it establishes a novel system of creation of common good that does social justice to its participants. Participation is voluntary, choice of manufacturers is free, rules are transparent and equally applied, but not too restrictive. Without the technological solutions presented in this book, establishing and automatically managing a scheme of such diversity and scale would be difficult, if not impossible.

The primary goal of the described scheme is to allow efficient traffic flow in the city and reduction of average travel times, number of taken rides, and gas emissions.

---

[1] https://www.theguardian.com/cities/2014/mar/20/licence-plate-driving-bans-paris-ineffective-air-pollution

[2] https://en.wikipedia.org/wiki/Tragedy_of_the_commons

But once the scheme is in place and perceived as stable and expected to last, it also opens up possibilities for the development of secondary business opportunities. In fact, the city should encourage secondary business value generation, as the two exist in a symbiosis which strengthens both parties – the city benefits by increasing the attractiveness of the scheme, the citizens by running or using new business opportunities. Concretely, the city could introduce emission/ride quotas, which could be traded among the citizens. This would make the scheme more attractive by allowing citizens that predominantly use public transport to sell their quotas and earn, while still maintaining the overall emission level within the projected limit. Businesses offering trading services would develop, where trading would surpass the mere exchange or payment for quotas.

The example with the smart-car controllers shows how a single scheme introduced by the city and embraced by the citizens through incentivization can produce various infrastructural, societal and business values. There is practically an unlimited number of possibilities for such multidimensional value generation that arise by managing physical, digital and human infrastructure in a coordinated way.

9.1.2 Making Citizens Active Smart City Stakeholders

Uber and other similar citizen-driver applications have broken the taboo of *citizen-provided services* and introduced the concept to the masses across the world. Even if the pioneering companies are often labeled as controversial or exploitatory and are met with protests and restrictions in various countries (especially where the activity in question is centrally controlled and tightly regulated) the idea of citizen-provided services has established itself as a viable reality. Private companies were the first to jump on the train and reap the benefits of the novel business and labor model, but the beneficiaries of it can be (and should be!) also the citizens and the municipalities. The key presumption for this is that the participation and profit-sharing rules are established by a trusted entity. In today's societies such entities are the institutions run by elected representatives. For this reason, (smart) cities are the perfect environments for establishing and developing these new labor and business models.

Citizen-provided services, due to their pervasiveness and responsiveness, can be a reasonable way to complement existing municipal inspection and maintenance services and reduce their overall costs. For example, take the globally popular bicycle-sharing schemes (e.g., Barcelona Bicing, Hangzhou Public Bicycle), where a city establishes a city-wide network of bicycle rental stations that can be used for a small fee. Maintaining such a network is extremely beneficial for a city, but at the same time also costly, as it requires constant inspection of bicycles and replenishment of stations. A common problem that occurs is that the stations on the periphery of the city quickly become empty in the morning, while those in the downtown/business parts get overfilled and lack empty parking slots. As a consequence, the unlucky user either has to lose time to find another station with available bicycles, or ride further away from the destination in search of a station with enough free slots. In the morning rush hour this can cause late arrival to work. In the afternoon, the situation is the opposite. This severely affects the usability of the scheme, as citizens cannot rely on the system and are unlikely to become regular users, defeating the original goal of the introduction of such scheme.

In order to keep the bicycle counts in balance and do the basic checking, the city has to deploy crews with trucks. Instead, the citizens themselves could be incentivized to re-arrange the bikes and report bikes with problems, leaving it to the municipality to do a single over-night tour of stations replacing the bicycles in disrepair. In order for the scheme to work, the incentives for transferring a bicycle have to be attractive and actively advertized to the nearby citizens. This means that the attractiveness of the reward has to be varied dynamically, increasing with the importance of the action (e.g., the reward should be higher if the station is already full than if a single slot is still empty). Lottery-like incentives (Chapter 8) are a good choice in this case as they allow the city to spend virtual lottery coupons instead of real money, and attract participation with actual rewards offered only after some period.

Once established, the bicycle-sharing scheme can be used as the infrastructural basis upon which to grow further business services. For example, under the assumption that the bicycles possess a transportation basket a citizen-run delivery service could be established, where citizens could be asked to transport a small (typically low value) shipment from A to B, where A and B lie on or near the route intended to be taken by the citizen, or run an errand.[3] In this case, the reward could be more substantial (e.g., monetary on the pay-per-performance basis) and proportional to the difficulty and amount of effort taken to perform the delivery. The reward would be split with the city for providing the physical infrastructure. While entrusting everyday citizens with commercial and legal responsibilities (in this case, for instance, associated with guaranteeing the delivery and privacy of the package) is a complex issue, this is something that we already do with services such as UberX (or People's Uber). Therefore, these are not obstacles that cannot be overcome with proper pre-screening, trust/reputation scores and signing of participation contracts beforehand. As more and more everyday shopping is being done online, existence of such a platform would be an efficient and ecological way to transport the goods quickly across the town for the low-value, low-priority delivery business segment, earning money for the city and the occasional citizen-cyclist, achieving a higher utilization of the existing cycling infrastructure, and if managed intelligently also allowing for a low-cost bicycle rearrangement across the stations.

Unlike Uber or ride-sharing schemes, here the citizens do not put at disposal their own physical infrastructure, but rather use the city's infrastructure (bicycles) to take part in business activities. This is considered the conventional approach, like when we use the city-built roads to do our business. What is different is that the citizens put their cognitive and physical capabilities (labor) at the disposal of other Smart City stakeholders (businesses). The synergy consists of businesses providing a business and profit opportunity, the city providing the infrastructure and the citizens providing the effort. All three stakeholders share the profits of the successfully

---

[3] www.taskrabbit.com, http://goget.my

accomplished task. What differentiates the described scheme and defines it as "smart" is the diversity and dynamicity. The tasks that need to be accomplished are not known in advance, the status of infrastructure changes dynamically (the station may be empty, a street closed) and the people providing the effort are not known in advance and execute the task with different quality of service.

The described scenario is a type of crowdsourced (or citizen-powered) maintenance. While managing the participants in this scenario is an extremely challenging task, the actual activities performed by the citizens are simple (inspecting a bicycle for basic functionality, traveling from station A to station B). This makes the scenario appear more plausible/feasible than if the activities of the individual participants involved some highly professional tasks, because we implicitly assume that a random person is capable of performing these simple tasks. However, from the management/controllability perspective the difference in complexity is minor. Managing ad hoc collectives of citizens that inspect and cater to cycling infrastructure has similar complexity as managing ad hoc collectives of engineers and technicians for predictive maintenance of air conditioning equipment. The experiences and technical know-how gathered through establishment of citizen-powered schemes such as the bike-sharing scheme can be further developed into different commercial or city-run schemes for predictive and corrective maintenance of equipment. In the Smart Cities of the future with millions of smart buildings equipped with all kinds of automated systems, sensors and IoT devices, the scale and pervasiveness of all this equipment makes it difficult to maintain.

This brings us to our next point – how to maintain the Smart City infrastructure?

9.1.3 Smart City Infrastructure Maintenance

Monitoring the diverse and pervasive equipment of a Smart City is an extremely complex research challenge. In this book we have presented a number of solutions that help us with this task by allowing us to abstract and uniformize the equipment, and perform remote maintenance operations from a logically centralized point of control. However, often remote (digital) maintenance (e.g., restarting a controller, recalibrating a sensor) is not enough, and additional physical intervention is required (e.g., to replace a part). Furthermore, due to the interaction between pieces of equipment produced by different manufacturers, in many cases a standardized maintenance procedure cannot be applied, and an ad hoc procedure must be worked out or approved by human experts. In such cases the interplay between human and software solutions is indispensable. The IoT Cloud infrastructure is responsible for monitoring and discovering potential problems, while teams of human professionals are engaged when needed, and possibly dispatched to perform maintenance activities on the spot.

The big players in the market have already moved into the area of professional human services. For example, both Amazon[4] and Google[5] already offer search and engagement of human professionals for various "home services," such as plumbing, A/C installation and electrical repairs.

At the moment, the platform is only the mediator between the service consumer and the service provider, vouching for the reputation of the service provider by performing a pre-screening of the candidates, and guaranteeing a refund in case of low service quality. It is not difficult to imagine, however, that future Smart City platforms could also take a more active role in managing the engagement of the professionals, in the way described in our motivating example. Having at its disposal a database of professionals with proven qualifications, experience and reputation, the platform can actively prompt the professionals to participate in ad hoc maintenance teams. The ultimate consumer in this case need not be the City (municipality) itself. The platform can again act as a trusted intermediary between commercial parties, facilitating the business transactions. This can prove to be an alternative to the current maintenance model where each smart building is connected to a static control center which centrally monitors and maintains all the devices, thus severely restricting the choice of device manufacturers and portability of devices.

The value such a scheme generates for the City is twofold. The diverse infrastructure of the city, both in public and in private ownership is monitored more closely and readily maintained, thus reducing costs in the longer run. Additionally, the platform actively promotes development and connection of local businesses with local customers. Instead of paying a costly service fee to a third-party smart-building-control center, the smart buildings and their residents are able to use the local professionals for most of the simpler maintenance work. This also allows small independent repair professionals to maintain their small businesses centered around local customers.

9.2 Towards the Smart Cities of the Future

Throughout this book we have presented a number of technological solutions that are the essential drivers and technological enablers of the novel Cyber-Human Smart City vision. But realizing the vision requires going beyond the technological advances to enable a holistic, sustainable ecosystem. The most important step in this process is abandoning the traditional, vertically closed Smart City and moving towards a horizontally integrated Cyber-Human Smart City (Section 9.1), which supports equal participation and integration of humans, smart devices and existing city facilities. The focus on the "historical verticals" [80] limits the innovation, sustainability and business potential of the city, thus its overall livability. Opening up this siloed view of the Smart City will allow more horizontal integration and will fuel the value generation process. Therefore, a Smart City needs to be a rich, self-sustaining

---

[4] http://marketingland.com/amazon-launches-home-services-potential-challenger-to-yelp-google-others-123460

[5] http://www.cnbc.com/2015/12/22/google-plays-matchmaker-expands-home-services-ads.html

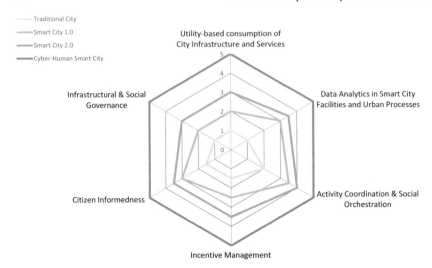

Fig. 9.1: Cyber-Human Cities road map

ecosystem that facilitates both production and consumption of added values for all the involved participants, ranging from humans to smart devices. The most significant advantage of this novel vision is clearly reflected in the structured and holistic value generation process, which is conceptualized in the proposed Architecture of Values (Chapter 1).

In order to help attain this goal, the presented technological solutions also need to be tightly horizontally integrated. The envisioned Smart City platform bears a comprehensive, integrative function, offering more than the mere "stitching together" of isolated technical solutions. Its primary goal is to facilitate the value generation process by providing the foundation for the development and hosting of integrative added-value services. A high-level overview of the Smart City platform is presented in Chapter 1. In the following, we mainly focus on the platform's core integrative functionalities. We discuss the the state-of-the-art solutions, their limitations, and identify future technology developments needed to realize Cyber-Human Smart Cities.

The role of ICT in Smart City value generation is not in doubt. However, a number of technical enablers are required to underpin this process. We have broadly categorized them in four main groups:

1) Provisioning and utility-based consumption of infrastructure and services
2) Incentive Management
3) Activity coordination & Social Orchestration
4) Monitoring & Data Analytics

Figure 9.1 shows a road map with necessary steps (Smart City maturity levels) towards realizing such an ecosystem for Cyber-Human Cities. The aforementioned enablers of the value generation process are some of the main dimensions of our road

	Infrastructure as Utility	Data analytics	Social Orchestration	Incentive Management	Governance	Citizen Informedness
ML-1	Stakeholders install, manage and consume exclusively their own smart devices	Facilities motoring data is collected manually. Little to no analytics	Manually-planned and orchestrated social orchestrations	Sparse or no use of incentives in digital domain	Traditional city governance policies are enforced "manually" through city's institutional mechanisms	No information personalization. Informedness through broadcasting or multicasting
ML-2	Smart City data can be shared among the stakeholders	Automatic data collection with offline mostly manual data analytics	Complex, small-scale static collaborations, planned and orchestrated by software	Heterogeneous, platform-specific, simple incentive schemes used independently by different organizations	E-governance and Open Data platforms	Information personalization for individuals Predefined delivery channels
ML-3	Common access to specific resources such as testbed facilities and living lab resources	Automatic data collection and near real-time data analytics	Large-scale, software-planned static collaborations on simple tasks (crowdsourcing)	More widespread use of incentives, limited to sharing economy, crowdsourcing. Simple schemes	Conceptual Smart City governance models and frameworks	Advanced information personalization. Delivery through multiple (IoT) channels
ML-4	Complete infrastructure is open and publicly accessible. Incentive mechanisms in place	Predictive, near real-time analytics spanning multiple city sectors	Small-scale, dynamically orchestrated complex collective collaborations.	Automated incentive management. More complex incentive schemes	Technology-enabled and automated governance of all Smart City facilities	Passive anticipated assistance and demonstrative use of IoT devices
ML-5	Smart City infrastructure is organized as a full-fledged public utility	Machine learning approaches with automated feedback loop	Large-scale, human-orchestrated, dynamic, complex, collective activities	Dynamic adaptations of incentives. Personalized, complex incentives for advanced collaborative tasks	Fully fledged socio-technical governance platform	Active anticipated assistance and demonstrative use of IoT devices coupled with incentives

Table 1: Summary of road map dimensions and their maturity levels

map for the development of a comprehensive Smart City platform. Additionally, we define two "cyber-human" maturity dimensions of the Smart City ecosystem:

5) Infrastructural & Social Governance
6) Citizen Informedness

These were identified as key enablers of horizontal integration and active citizen involvement in Smart Cities of the future. Progressing along these dimensions are the necessary steps to shift from traditional Smart Cities towards future Cyber-Human Cities. In a sense, they can be seen as different maturity levels of the Cyber-Human Cities. In the remainder of this section we discuss the most important dimensions of the road map in more detail.

9.2.1 Utility-Based Consumption of Smart City Infrastructure and Services

One of the dimensions (Fig. 9.1) of the Cyber-Human City platform road map is the level to which it allows organization of Smart City ICT infrastructure as a publicly accessible and open utility. This is reflected in its support for *utility-based consumption/delivery of Smart City infrastructure and services*, in the sense of the extent to which Smart City stakeholders are enabled to engage in utility generation and consumption, as well as its distribution. We have identified the following maturity levels (ML) of utility-based consumption of Smart City ICT infrastructure and services: ML-1) Traditional consumption, where each citizen or organization installs, manages and consumes exclusively their own smart devices; ML-2) The devices are still managed in the traditional manner, but the data can be shared among the stakeholders, e.g., Xively; ML-3) Common access to specific resources such as testbed facilities and living lab resources, e.g., Living Labs [159]; ML-4) Complete Smart City ICT infrastructure (e.g., sensory data, smart devices, smart home gateways, etc.) is open and publicly accessible. Incentive mechanisms are implemented to stimulate and facilitate infrastructure sharing; ML-5) Smart City infrastructure is organized as a public utility. Delivery-consumption-compensation models are developed, enabling Smart City stakeholders to engage in utility generation and consumption, as well as its distribution in a self-regulating market. From Fig. 9.1, we notice that traditional cities are at ML-1, Smart Cities 1.0 are at ML-2 and Smart Cities 2.0 are at ML-3. Finally, to realize Cyber-Human cities an ML-5 of consumption of city infrastructure and services is required.

The current state of the art can bring Smart Cities to ML-3 with respect to this dimension. By integrating the solutions proposed in Chapters 2, 3 and 6 the Cyber-Human platform is mature enough to facilitate open access to the Smart City infrastructure resources and provide foundational automated incentive management mechanisms towards a city-wide market for sharing infrastructure resources. By integrating the proposed solutions to advance the Cyber-Human platform in this direction, we can effectively enable Smart Cities to evolve to ML-4.

However, our vision of the Cyber-Human City requires a full-fledged organization of Smart City infrastructure as a public utility (ML-5 in Fig. 9.1). One of the key preconditions is to provide novel support for realizing the delivery-consumption-compensation model for the infrastructure resources. Traditional public utilities exclusively rely on existing markets, business models and monetary institutions to realize this model. However, to realize broader participation in the previously presented architecture of values Smart Cities largely lack suitable business models for exchanging resources and services among the stakeholders. Moreover, infrastructure owners and infrastructure brokers require an ecosystem to support trading of Smart City services and assets. Unfortunately, this is largely missing in current Smart City platforms.

Traditional models, e.g., banking/payment-processing systems, fall short regarding the speed, scale and agility required to support trading in our Smart City ecosystem: i) They mainly rely on invoicing as the only means to perform a monetary transaction. ii) Banks only do business with people, not smart devices (which are active participants in the Smart City platform). iii) privacy issues when trading with sensitive information due to the involvement of a third party, e.g., a bank. iv) There is a fixed lower boundary for a transaction amount, e.g., 0.01 EUR. v) Duration of asset transfer or legal boundaries.

Therefore, to realize the utility-oriented delivery and consumption of Smart City infrastructure resources and reach ML-5, we need to extend current models and solutions to support the following set of design principles: *Smart City trader units* – Devices and services/applications autonomously decide with whom to trade and do business; *Automated cash handling* – It is difficult to manage and oversee individual devices in the large-scale hyper-distributed environments, thus devices need to have higher degree of autonomy; *Micro-transactions* (time- and size-wise) –

Enabling pay-as-you-go consumption of IoT infrastructure (e.g., per data instance) with small/no transaction fees across regions with different location and compliance criteria; *Scalable transaction processing* – Supporting the large number of devices, e.g., gateways capable of providing resources/capabilities and performing business transactions; *No central authority* – Brings considerable benefits for the privacy requirements, but keeps the whole process highly transparent.

One promising technology that can be used as a base for the solutions that can support these principles is Blockchain [117, 141, 146]. Generally, a Blockchain is a distributed database that maintains a continuously growing list of data records. Each block holds a batch of transactions and since it is based on P2P consensus and strong encryption it is very resilient against tampering and revision. This makes it a good solution for any kind of transactions within inherently untrusted IoT networks. For example, it could be used by smart devices to autonomously trade resources, e.g., sensory data, storage and network capacity among themselves, but also for secure file transfer or different kinds of user-defined smart contracts. However, one of the limitations of current Blockchain solutions is their lack of scalability, and although there are partial solutions (e.g., side chains) a more scalable approach is required to accommodate the number and frequency of transactions envisioned in our future Smart City.

9.2.2 Data Analytics in Smart City Facilities and Urban Processes

The core of any Smart City is the ability to monitor various city facilities in order to analyze their performance and optimize urban processes, e.g., within a specific sector. Data analytics is probably the most advanced field in contemporary Smart Cities and today it is usually used as a synonym for "smartness" in Smart Cities, mainly because it can support answering different policy, planning, governance and business questions, support decision making in enabling a smarter environment and even allow for automated predictions and recommendations about urban processes. Despite this there remain challenges to enable the horizontal (across different city sectors) integration of Smart City data, which would allow full utilization of the potential of Smart City data analytics. Our road map caters to this need by defining a dimension that is used to evaluate the maturity level of *Data analytics in Smart City facilities and urban processes*. As illustrated in Fig. 9.1, we define the following maturity levels (ML) of Smart City data analytics: ML-1) Facilities monitoring data is collected manually and there is little to no analytics of facilities performance, e.g., taking a reading from an energy meter. ML-2) Automatic data collection with offline mainly manual (software-assisted) data analytics of the facilities; ML-3) Automatic data collection and near real-time data analytics, based on traditional models, e.g., energy consumption models. ML-4) Predictive near real-time analytics for predictions and recommendations, potentially across multiple city sectors, e.g., energy management and building management. ML-5) Machine learning approaches with automated feedback loop that enable continuous, fully automated optimization of facilities (via actuators), e.g., energy distribution and load balancing. From Fig. 9.1, we can observe that traditional cities are at ML-1 and Smart Cities 1.0 are at ML-3. Current developments in data analytics are enabling rapid evolution of Smart Cities towards ML-4, i.e., enabling cross-domain predictive data analytics. Currently we are working on addressing the challenges of reaching ML-4 of data analytics in Smart Cities in the scope of the SMART-FI[6] project.

The main objective of our current research effort in Smart City data analytics is to facilitate the interaction between humans and smart devices in three main activities: collecting, communicating and exploiting Smart City data. To this end, we are developing a platform that strives to enable collection of the data from a variety of sources, such as sensors, mobile devices or public open data services. Secondly, the platform will provide mechanisms to enable the data coming from different networks and protocols to be homogenized and communicated. Finally, it will provide facilities to develop, deploy and orchestrate novel Smart City services, e.g., for predictions and recommendations, in order to enable the city data to be transformed into disruptive innovation building blocks for the Smart Cities of the future.

The main features of the aforementioned platform include design methodologies and models for implementing unified Smart City data analytic functions, as well as the runtime models for integration, deployment and operation of Smart City services. These models are intended to enable seamless development and management of micro data analytics services, which will serve as one of the main building blocks for sustainable data analytics in future Smart Cities. We are also building a comprehensive tool suite, based on cutting-edge data-processing architectures and technologies such as lambda architecture, microservices and reactive data streams. The main objective is to address current challenges in Smart City data analytics, such as dynamic, on-demand governance and elasticity management of the resource, quality and cost dimensions of the data analytics micro services.

Finally, as shown in Fig. 9.1, to fully realize our vision of Cyber-Human Cities, we need to advance contemporary data analytics approaches to the next level (ML-5). This requires support for automated, online learning algorithms, based on deep-learning neural networks in the context of Smart Cities, as well as provision of mechanisms to support city-scale automated feedback loops in order to enable continuous, fully automated monitoring, analytics and optimization of Smart City facilities and processes. Although there are approaches based on machine learning and data mining (e.g., [32, 179, 149]), which deal with Smart City analytics, a number of challenges remain to enable the development of Smart City data analytics models that exceed simplistic statistical models in complexity. Final precondition for reaching ML-5 is support for continuous feedback loops, especially remote, real-time actuation in Smart Cities. In our previous work, we have developed a comprehensive programming model for developing large-scale IoT Cloud systems [121, 124, 122]. This model can serve as a solid foundation for realizing Smart City feedback loops, however additional work is required to overcome the challenges of integrating such IoT

---

[6] http://smart-fi.eu

Cloud applications with machine learning approaches in a scalable and sustainable manner.

9.2.3 Social Orchestration & Activity Coordination – Supporting Complex Coordinated Activities

Unlike the general topic of activity coordination and composition[7] in computer science, which deals with the coordination of software processes, agents and computing nodes, the concern of Social Orchestration is the coordination and composition of activities collectively performed by the human participants. Both are extremely important in the context of the Cyber-Human Smart City. However, in this chapter the focus is placed on Social Orchestration, as a novel and still underdeveloped area that underpins the citizen inclusion in Smart City processes. Concretely, we discuss the necessary technical solutions for successfully including and managing humans in collective activities in a Smart City environment, and define the maturity levels to help identify particular development stages of Social Orchestration support.

Social orchestration, even when not explicitly singled out as a distinctive feature of a system, is a core functionality of any social-computing, socio-technical, crowdsourcing or human-based CAS platform. In Section 6.1 we gave an overview of state-of-the-art social computing systems and research prototypes. Based on this overview, we are able to formulate the following distinctive properties that characterize social orchestration.

Orchestration Scale. The order of magnitude of the number of collaboration participants. It can vary from an individual performing a single independent activity (e.g., answering a question) to a collaborative activity involving a significant proportion of the overall city population (e.g., participating in an election).

Task Complexity. Defining what it means for a collective activity (task) to be complex is highly context-specific. At the most general level, complex tasks are those that can be broken down into a number of simpler (atomic) subtasks, possibly all the way down to the subtasks that can be performed by a single person. The subtasks can have logical or temporal interdependencies. The more interdependencies there are, the more complex the aggregate task is. Task complexity can be considered from two aspects: 1) Planning – how difficult it is to break down (analyze) a complex task into a sequence of optimal subtasks; and 2) How difficult it is to execute the already planed task, i.e., execute the subtasks.

Planning. Planning can be performed by software, by humans, or hybridly. Planning the execution of a complex task is often an algorithmically hard problem, unsuitable for humans. When the task is a standardized one, for which known planning solutions exist, we often use an out-of-the-box software planner, or develop one if the savings obtained through the optimality are worth it. We encounter such

---

[7] See [2], Section III.7 for an overview of the area.

tasks mainly in well-established and frequently repeated business processes (e.g., transportation, chain-supply).

In social environments, however, the tasks are rarely standard or even known in advance, meaning that the likelihood of having at our disposal a ready-made planning algorithm is low. At the same time, in such environments it is usually not time- or cost-effective to develop scenario-specific algorithms. Furthermore, such tasks are rarely dependent on the optimality of the solution, and in many such cases where the optimality/accuracy can be sacrificed for speed, humans might be able to provide acceptable ad hoc heuristics. Consider, for example, a user seeking to solve a real-life situation similar to a river-crossing problem[8] with the help of the Smart City platform. While such problems have optimal solutions, in a real-life situation adapting an existing algorithm to provide the optimal solution in the given situation is too slow, while crowdsourcing such a problem is able to quickly provide a solution [143]. Of course, human-based planning is far from trivial, as it often requires complex negotiation, agreement and aggregation algorithms (cf. [148, 102]). The results, however, are applicable to a broader class of problems.

Finally, hybrid planning means that both algorithmic and human capabilities are applied to construct a plan. Often, the human component is used for interpreting the context properly, and selecting/adapting an existing suitable algorithm or software tool to perform the planning.

Dynamicity. This property refers to the time when a plan is generated/adapted. If the planner possesses all the necessary information (inputs) to construct a plan before the actual execution we refer to such execution as static. If the plan cannot be fully constructed without inputs that are unknown until runtime, we refer to such execution as dynamic. Most state-of-the-art social-computing platforms currently only support static execution, with notable research prototype exceptions, such as CrowdLang [112] and Smart Society [154], which demonstrate different levels of execution dynamicity.

We can now define the maturity levels as follows: ML-1) Manually planned and orchestrated small-scale social orchestrations, e.g., management of rescuers in case of an accident; ML-2) Complex but small-scale static collaborations, typically business processes, planned and orchestrated by software. Example: various BPEL orchestration engines[9] with support for BPEL4People; ML-3) Widespread use of large-scale static collaborations, for simple tasks with respect to both planning and execution. Planning done by software; exceptionally also "manually" by human individuals. Example: existing crowdsourcing platforms [44]; ML-4) Small-scale complex collaborations allow creation of ad hoc teams (collectives) of professionals. The planning and execution is mostly software-driven, but to a limited extent, humans are able to influence parts of the execution at runtime; ML-5) Full-scale negotiations and agreements of human participants dynamically determine the further execution course (dynamic human planning). Human planners make extensive use of software

---

[8] https://en.wikipedia.org/wiki/River_crossing_puzzle

[9] https://en.wikipedia.org/wiki/List_of_BPEL_engines

services for planning. Collaborations are large scale, involving entire neighborhoods or population groups.

ML-2 level corresponds to a traditional city, which extensively uses manual social orchestration in non-business environments, and often a software-orchestrated static orchestration in business processes. A Smart City 1.0 with its crowdsourcing and resource pooling fulfills the maturity criteria of ML-3, and slightly surpasses it when considering the recent onset of augmented-reality games such as Pokemon Go, which allow humans to influence the orchestration and organize collectives. It is to be expected that this trend will further develop in the Smart City 2.0 which will likely exhibit all the complexity of level ML-4. ML-5 corresponds to the Cyber-Human Smart City vision. The framework for programmatic coordination and collaboration management presented in Chapter 7 is the direct enabler of ML-4 and sets the technological foundation for ML-5.

9.2.4 Incentive Management

As we have argued extensively in Chapter 8, incentives are an indispensable enabler and a control mechanism of social orchestrations. Therefore, any Smart City attempting to involve citizens more deeply and more actively will need to use some kind of incentives to help foster participation, minimize attrition rates, improve quality of contributions and combat dysfunctional behavior. However, only recently has awareness of the importance of incentives and incentive management started to grow significantly. The result is that advances in this area are still lagging behind developments in the other technical "dimensions," but they look set to rapidly close the gap.

When comparing the levels of complexity and advancement of incentive management, we observe the following properties.

Automated Incentive Management. Allows specification of complex incentive mechanisms from standardized (library) components and automated application of the mechanisms.

Incentive Comparability. The adoption of Automated Incentive Management solutions is a technical enabler for the subsequent adoption of similar (comparable, standardized) incentive mechanisms across different Smart City platform applications. This allows citizens to assess the potential benefits among different incentive schemes, increasing the overall transparency of the added-value services of the Smart City platform.

Reputation Transfer. Standardized rewards and achievements are transferable, allowing citizens to skip the cold-start problem and be selected to perform tasks where a specific degree of reputation or qualifications are required. This is a necessary precondition for sustainable digital careers, it increases citizen mobility, and it fosters the generation of novel business values.

Complexity. Refers to the number of employed incentive mechanisms in a single incentive scheme. Powerful incentive schemes typically combine a number of incentive mechanisms targeting different behavioral aspects (Chapter 8).

Dynamicity. Traditionally, incentive mechanisms are static rules that are only manually (and relatively rarely) changed. In contrast, incentive mechanisms can also be dynamically generated or adapted, i.e., constantly re-tailored to the targeted citizen group or to the digital artifact with which they are associated [157].

We can now define the maturity levels as follows: ML-1) Incentives in the digital domain are not used, or used very sparsely; ML-2) Incentives are used rarely and independently by different organizations and platforms. Incentives are heterogeneous and platform-specific and applied to users of those platforms/organizations only. Mostly simple incentive schemes. No automated incentive management. No incentive comparability; ML-3) Incentives are more commonly used but mostly to stimulate sharing of resources (computing resources, devices, joint activities such as ride-sharing) or simple crowdsourcing, but not of advanced cognitive/physical labor nor collective complex activities. Incentive schemes are predominantly simple and non-personalized; ML-4) Automated incentive management is used to produce concrete incentives out of general incentives. This allows the use of more complex incentive schemes. Incentives are more standardized and uniform, comparable across different platforms; ML-5) Incentive mechanisms are dynamically generated and adjusted, tailored to the targeted user group or the artifact with which they are associated [157]. Reputation transfer is possible.

Many traditional cities, especially in less developed countries, have yet to start using incentives in the digital domain. While some smaller innovative companies may be using them, none of the city services are using them. The reason is that the sense that the city administration and infrastructure is there to serve its citizens is lacking, so there is no push to try to additionally motivate the citizens to get actively included in the functioning of the city. This places such cities on level ML-1. On the other hand, some cities that host a technologically advanced local economy or have an increased awareness of the importance of citizen inclusion are exhibiting all the characteristics of level ML-2. A Smart City 1.0 is predominantly focused on the usage optimization of infrastructural resources, and thus approaching maturity level ML-3. A Smart City 2.0 is starting to pay more attention to the human capital of the city. It is therefore realistic to expect that the accepted understanding of such a city will exhibit many of the characteristics of level ML-4. Level ML-5 corresponds to the described Cyber-Human Smart City vision. Incentive management technology, such as the PRINGL framework presented in Chapter 8, is one the necessary enablers of maturity levels ML-4 and ML-5.

9.2.5 Citizen Informedness

In Section 1.6 we wrote about the education, informedness and active inclusion of citizens in different life aspects of a city as the fundamental enablers of citizen empowerment – one of the defining characteristics of a Cyber-Human Smart City.

Before defining the maturity levels, we must first define more precisely the different characteristics/functionalities that a Cyber-Human Smart City platform can implement in order to support different aspects of citizen informedness.

Delivery through IoT environment. The pervasiveness of the devices and their presence in everyday objects makes them familiar and trusted. Therefore they are a suitable medium for transmitting knowledge and information in small, easily absorbable quantities. This can include advice coming from the devices in a smart home about the amount of electricity the vacuum cleaner has used, or tips on energy savings. In addition, since the incentives need to be advertised to the potential users, the same environment can be used to notify the citizens of the benefits of participation in specific city-run schemes.

Learning by example. Whenever applicable, practical demonstration of the usage should be the favored approach. This concept is applicable at all scales, ranging from the use of a single smart device, to a smart home, to the testing of city-wide changes (such as changes in traffic regulation).

Information personalization. The Smart City platform services should filter and deliver to the users (citizens) personalized information based on learned behavior patterns and scheduled events. The concepts of IoT delivery and information person-alization are already being practically used in commercial products such as Amazon Echo and Google Home.

Anticipated assistance. The Smart City platform services or personal IoT devices should act as autonomous agents in good faith on the citizen's behalf and engage in benevolent interactions with other citizens' devices and services in order to increase potential benefits. The citizen decides the agent's allowed level of autonomy and personally authorizes all legally binding activities. Smart City trader units are a good example of how this concept can be applied to empower the citizens to trade personal resources and consume IoT infrastructure as a utility.

We can now define the maturity levels as follows: ML-1) Complete lack of infor-mation personalization. Citizens are informed through broadcasting (TV, newspapers) or multicasting (mass SMS/email/leaflet campaigns to population groups) without individual discrimination; ML-2) Information personalization is used to provide information of interest to each individual citizen. The delivery channel is typically a single, predefined one – mostly email, or a smartphone application; ML-3) Informa-tion personalization is more advanced and delivered through a variety of channels spanning various IoT devices in a citizen's environment; ML-4) The IoT devices are used to demonstrate and simulate different usage scenarios but on the individual user level (e.g., to teach the user how to use a device). Anticipated assistance is used, but only passively to filter the data, not to initiate interactions on the user's behalf (such as micro-transactions and barters); ML-5) The Smart City platform orchestrates a large number of IoT devices and administers incentives to large groups of citizens to simulate potential outcomes of collective decisions (e.g., simulating the outcomes of a political decision). Anticipated assistance is widely used to relieve citizens of taking part in many SmartCity interactions. The software assistants are fully autonomous within the given limits, and engage in trading and other interactions on the user's behalf (e.g., autonomously reserving a trip with accommodation on the user's behalf or commercially leasing the user's currently free computing resources to a third party).

Traditional cities are currently located between levels ML-1 and ML-2 – information personalization is used, but mostly by private companies as a means of increasing the attractiveness of their own services (Google Now, Apple Siri). The city itself does not usually target its citizens personally. A conventional Smart City 1.0 is at ML-3 – the IoT environment is used here for displaying and collecting information. ML-4 level currently represents the state of the art in research (e.g., Living Labs [159], autonomous negotiation agents [135]), but no significant practical/commercial applications exist. ML-5 level represents the ultimate stage envisioned by the Cyber-Human Smart City, where dealing with large amounts of low-level information is delegated to software agents, leaving more time for the citizens to engage in creative, political or relaxing activities.

9.2.6 Infrastructural and Social Governance

Automated governance is poised to become one of the cornerstones of Smart City management. There are two diversity aspects that governance needs to cover: *vertical* – allowing the transformation and enforcement of high-level city policies and goals via a number of low-level operational procedures; and *horizontal* – allowing the application of operational procedures over various infrastructural and social resources and stakeholders with different properties, objectives, interests and backgrounds. This diversity calls for the automation of the whole process, which inevitably must consider hardware, software and humans and their interplay in the context of a Smart City. As such, diversity management will play an important role in future Smart City platforms, constituting the city's *Administrative Infrastructure* (see Figure 1.2).

We have identified the following maturity levels of Smart City governance: ML-1) Traditional city governance without ICT support. Governance targets the physical infrastructure management processes under the city's control. The governance enforcement is performed "manually" through the city's institutional mechanisms; ML-2) E-governance and Open Data which assume integration of various traditional stand-alone systems and services between government-to-customer (G2C), government-to-business (G2B) and government-to-government (G2G) [18]; ML-3) Conceptual Smart City governance models and frameworks, mainly focusing on high-level Smart City objectives and policies; ML-4) Technology-enabled and automated Smart City governance, enabling fine-grained and logically centralized control of geo-distributed Smart City ICT infrastructure and applications; ML-5) Social governance. In addition to the management of ICT infrastructure, the social infrastructure is also automatically managed, i.e., the platform supports procedures and virtual organizations catering to the social needs of participating citizens. In Figure 9.1, the *Infrastructural & Social Governance* dimension illustrates the described governance road map in relation to other dimensions.

As we have discussed in Section 9.1, ICT applications and services are becoming an integration factor in optimizing urban processes, infrastructure and facilities, such as urban transportation and energy management. In Chapter 5, we have discussed the current ICT governance challenges in Smart Cities and proposed a governance framework that is capable of supporting ML-4 of Smart City governance. We have shown how this approach can be used to bridge the gap between high-level governance objectives (which mainly concern city representatives and business stakeholders) and operations processes. The latter concern technical stakeholders such as Smart City service developers and operations managers, who need to implement concrete operations processes conforming to or enforcing the high-level governance objectives. We have demonstrated how our approach facilitates automated mapping and enforcement of the high-level policies down to the technical level, i.e., Smart City infrastructure and applications, effectively decreasing the risk of lost requirements or over-regulated systems, reducing management costs and nourishing innovation opportunities.

Apart from some work [81, 164] and initiatives[10] on digital labor organization and sustainability, the field of social governance has so far remained tucked away in purely academic discourse. However, in Cyber-Human Cities automating also the social governance is considered as one of the key goals that need to be achieved in order to reach ML-5 in Fig. 9.1. At this level the Smart City platform is expected to provide technological tooling or environments for unsupervised interaction and exchange of opinions among participants in collective activities (virtual social fabric), mechanisms for ensuring/certifying transparency and fairness of all employed algorithms governing citizen participation, and established procedures for filing complaints or addressing unjust treatment. Digital (self-)organizations, such as digital unions or interest groups, should provide an institutional means of representation and a tool for fighting for own interests.

9.3 Conclusion

In this chapter we have presented the necessary requirements and a road map with concrete technological advancements needed to move beyond contemporary Smart Cities and towards the Smart Cities of the future. We have discussed how horizontal integration across the variety of Smart City sectors and domains can serve as the main driver behind the value generation process in the architecture of values. To facilitate such a horizontal integration, we proposed empowering and exploiting in synergy diverse Smart City stakeholders. This is in contrast with the conventional

---

[10] http://www.faircrowdwork.org/

"representative-driven" development approaches of existing Smart City initiatives. Based on a number of concrete application examples, we argued that making citizens active stakeholders can bring a range of benefits such as first-hand experience, overall presence in space and time, and better technical know-how, e.g., gathered through establishment of citizen-powered schemes. Another key accelerator of value generation is the structured provisioning and governance of Smart City infrastructure and applications based on comprehensive automation tools. We posit that an efficient and effective management of Smart City systems is the key stepping stone towards utility-based consumption of Smart City resources and more importantly a crucial step towards democratizing Smart City facilities and values.

Based on these observations, requirements, our previous work and state-of-the-art approaches we have derived a comprehensive road map to realize the presented vision of Cyber-Human Smart Cities. The proposed road map describes the expected future advances, in terms of novel models and technologies, that are needed to foster and fully utilize the value generation process in Cyber-Human Cities. Since Smart Cities are a complex, living and ever-evolving ecosystem, we have structured the road map along six fundamental dimensions, each defining clear maturity levels. Although the advances along the individual dimensions undoubtedly bring additional benefits to all the involved Smart City stakeholders, only by adopting a holistic approach that pushes forward along all the dimensions and blends in the proposed concepts, models and technologies can the full potential of future Cyber-Human Cities be realized.

As a final word, we hope that this book can serve as a signpost indicating future directions in the multidisciplinary Smart City research area, but also offer a preview of the upcoming challenges and required development activities to Smart City managing authorities and interested business stakeholders.

References

1. Skinner, b. f. science and human behavior. new york: The macmillan company, 1953. 461 p. Science Education **38**(5), 436–436 (1954). DOI 10.1002/sce.37303805120. URL http://dx.doi.org/10.1002/sce.37303805120

2. Smartsociety consortium, deliverable 1.1 - interdisciplinary foundations of smart societies. http://www.smart-society-project.eu/publications/deliverables/D_1_1 (2013)

3. Smartsociety consortium, deliverable 5.3 - specification of advanced incentive design and decision-assisting algorithms for cas. http://www.smart-society-project.eu/publications/deliverables/D_5_3 (2015)

4. Adar, E.: Why i hate mechanical turk research (and workshops). In: Proc. of CHI'11 Workshop on Crowdsourcing and Human Comp. ACM, Vancouver, Canada (2011)

5. Ahern, D.M., Clouse, A., Turner, R.: CMMI distilled: a practical introduction to integrated process improvement. Addison-Wesley Professional (2004)

6. Ahmad, S., Battle, A., Malkani, Z., Kamvar, S.: The jabberwocky programming environment for structured social computing. In: Proc. 24th Annual ACM Symposium on User Interface Software and Technology, UIST '11, pp. 53–64. ACM (2011). DOI 10.1145/2047196.2047203

7. Alam, S., Chowdhury, M., Noll, J.: Senaas: An event-driven sensor virtualization approach for internet of things cloud. In: NESEA (2010)

8. Armbrust, M., Fox, A., Griffith, R., Joseph, A.D., Katz, R., Konwinski, A., Lee, G., Patterson, D., Rabkin, A., Stoica, I., et al.: A view of cloud computing. Communications of the ACM **53**(4), 50–58 (2010)

9. Armstrong, M.: Armstrong's Handbook of Reward Management Practice: Improving Performance Through Reward, 3rd edn. Kogan Page Publishers, London (2010)

10. Armstrong, T., Trescases, O., Amza, C., de Lara, E.: Efficient and transparent dynamic content updates for mobile clients. In: Proceedings of the 4th international conference on Mobile systems, applications and services, pp. 56–68. ACM (2006)

11. AWS: CloudFormation. URL: https://aws.amazon.com/cloudformation/. [Online; accessed Feb.-2015]

12. Bano, M., Zowghi, D., Ikram, N.: Alignment between business requirements and services: the state of the practice. In: ICSSEA (2013)

13. Bardin, J., Lalanda, P., Escoffier, C.: Towards an automatic integration of heterogeneous services and devices. In: APSCC, pp. 171–178 (2010)

14. Baresi, L., Heckel, R.: Tutorial introduction to graph transformation: A software engineering perspective. In: A. Corradini, H. Ehrig, H.J. Kreowski, G. Rozenberg (eds.) Graph Transformation, *LNCS*, vol. 2505, pp. 402–429. Springer (2002). DOI 10.1007/3-540-45832-8_30

15. Barowy, D.W., Curtsinger, C., Berger, E.D., McGregor, A.: Automan: A platform for integrating human-based and digital computation. SIGPLAN Not. **47**(10), 639–654 (2012)

© Springer International Publishing AG 2017
S. Dustdar et al., *Smart Cities*,
DOI 10.1007/978-3-319-60030-7

16. Bloom, M., Milkovich, G.: The relationship between risk, incentive pay, and organizational performance. The Academy of Management J. **41**(3), 283–297 (1998)
17. Bonomi, F., Milito, R., Zhu, J., Addepalli, S.: Fog computing and its role in the Internet of Things. In: MCC workshop on Mobile cloud computing, pp. 13–16 (2012)
18. Bose, S., Rashel, M.R.: Implementing e-governance using oecd model (modified) and gartner model (modified) upon agriculture of bangladesh. In: Computer and information technology, 2007. iccit 2007. 10th international conference on, pp. 1–5. IEEE (2007)
19. BOSH: BOSH. URL: http://docs.cloudfoundry.org/bosh/. [Online; accessed Feb.-2015]
20. Bowerman, B., Braverman, J., Taylor, J., Todosow, H., Von Wimmersperg, U.: The vision of a smart city. In: 2nd International Life Extension Technology Workshop, Paris, vol. 28 (2000)
21. Bozzon, A., Brambilla, M., Ceri, S., Mauri, A., Volonterio, R.: Pattern-based specification of crowdsourcing applications. In: Proc. 14th Intl. Conf. on Web Engineering (ICWE) 2014, pp. 218–235 (2014)
22. Bozzon, A., Fraternali, P., Galli, L., Karam, R.: Modeling crowdsourcing scenarios in socially-enabled human computation applications. Journal on Data Semantics pp. 1–20 (2013)
23. BusyBox: BusyBox: The Swiss Army Knife of Embedded Linux. URL: https://busybox.net/about.html. [Online; accessed Jan.-2015]
24. Buyya, R., Yeo, C.S., Venugopal, S., Broberg, J., Brandic, I.: Cloud computing and emerging it platforms: Vision, hype, and reality for delivering computing as the 5th utility. Future Generation computer systems **25**(6), 599–616 (2009)
25. Candra, M., Truong, H.L., Dustdar, S.: Provisioning quality-aware social compute units in the cloud. In: S. Basu, C. Pautasso, L. Zhang, X. Fu (eds.) Service-Oriented Computing, *Lecture Notes in Computer Science*, vol. 8274, pp. 313–327. Springer Berlin Heidelberg (2013). DOI 10.1007/978-3-642-45005-1_22. URL http://dx.doi.org/10.1007/978-3-642-45005-1_22
26. carriots.com: Carriots–IoT Application Platform. URL: https://www.carriots.com. [Online; accessed Jan.-2015]
27. Casati, F., Daniel, F., Dantchev, G., Eriksson, J., Finne, N., Karnouskos, S., Montera, P.M., Mottola, L., Oppermann, F.J., Picco, G.P.: Towards business processes orchestrating the physical enterprise with wireless sensor networks. In: ICSE'12, pp. 1357–1360 (2012)
28. Charfi, A., Mezini, M.: Hybrid web service composition: business processes meet business rules. In: ICSOC, pp. 30–38. ACM (2004)
29. Chenu-Abente, R., Maltese, V., Hume, A., Kharkevich, U., Fischer-Hübner, S., Martucci, L.A.: Smartsociety consortium, deliverable 4.2 - peer search in smart societies. http://www.smart-society-project.eu/publications/deliverables/D_4_2
30. Chun, B.G., Ihm, S., Maniatis, P., Naik, M., Patti, A.: Clonecloud: elastic execution between mobile device and cloud. In: Conference on Computer systems. ACM (2011)
31. Ciciriello, P., Mottola, L., Picco, G.P.: Building virtual sensors and actuators over logical neighborhoods. In: International workshop on Middleware for sensor networks, pp. 19–24. ACM (2006)
32. CITIES, S.: Trace analysis and mining for smart cities: issues, methods, and applications. IEEE Communications Magazine **121** (2013)
33. Copie, A., Fortis, T., Munteanu, V.I., Negru, V.: From cloud governance to iot governance. In: Advanced Information Networking and Applications Workshops, pp. 1229–1234. IEEE (2013)
34. CoreOs: CoreOS - a Linux for Massive Server Deployments. URL: http://coreos.com/. [Online; accessed Mar.-2016]
35. Cuervo, E., Balasubramanian, A., Cho, D.k., Wolman, A., Saroiu, S., Chandra, R., Bahl, P.: Maui: making smartphones last longer with code offload. In: Proceedings of the 8th international conference on Mobile systems, applications, and services, pp. 49–62. ACM (2010)
36. Davidson, Emily A (Softchoice Advisor): The Software-Defined-Data-Center (SDDC): Concept Or Reality? URL: http://tinyurl.com/omhmbfv. [Online; accessed Jan.-'15]

37. Dawson-Haggerty, S., Jiang, X., Tolle, G., Ortiz, J., Culler, D.: smap: a simple measurement and actuation profile for physical information. In: SenSys, pp. 197–210 (2010)
38. De Souza, L.M.S., Spiess, P., Guinard, D., Köhler, M., Karnouskos, S., Savio, D.: Socrades: A web service based shop floor integration infrastructure. In: The internet of things, pp. 50–67 (2008)
39. Deci, E., Ryan, R.: Intrinsic Motivation and Self-Determination in Human Behavior. Plenum Press (1985)
40. Diller, J., Song, S.: Method of and a system for ranking members within a services exchange medium (2014). URL http://www.google.com/patents/US8700614. US Patent 8,700,614
41. Diochnos, D.I., Rovatsos, M.: Smartsociety consortium, deliverable 6.2 - static social orchestration: implementation and evaluation. http://www.smart-society-project.eu/publications/deliverables/D_6_2/ (2015)
42. Distefano, S., Merlino, G., Puliafito, A.: Sensing and actuation as a service: a new development for clouds. In: NCA, pp. 272–275 (2012)
43. Distefano, S., Merlino, G., Puliafito, A.: A utility paradigm for IoT: The sensing Cloud. Pervasive and mobile computing **20**, 127–144 (2015)
44. Doan, A., Ramakrishnan, R., Halevy, A.Y.: Crowdsourcing systems on the world-wide web. Comm. ACM **54**(4), 86–96 (2011). DOI 10.1145/1924421.1924442
45. Don DeLoach: Internet of Things Part 4: Critical issues around governance for the Internet of Things. URL http://tinyurl.com/mxnq3ma. [Online; accessed July-2014]
46. Dorn, C., Edwards, G., Medvidovic, N.: Analyzing design tradeoffs in large-scale sociotechnical systems through simulation of dynamic collaboration patterns. In: R. Meersman, H. Panetto, T.S. Dillon, S. Rinderle-Ma, P. Dadam, X. Zhou, S. Pearson, A. Ferscha, S. Bergamaschi, I.F. Cruz (eds.) OTM Conferences (1), *Lecture Notes in Computer Science*, vol. 7565, pp. 362–379. Springer (2012)
47. Dustdar, S., Guo, Y., Satzger, B., Truong, H.L.: Principles of elastic processes. Internet Computing, IEEE **15**(5), 66–71 (2011)
48. European Commission: Report on the Consultation on IoT Governance. URL: http://tinyurl.com/mx24d9o. [Online; accessed August-2014]
49. European Commission: Report on the public consultation on IoT governance. URL: http://tinyurl.com/mx24d9o. [Online; accessed August-2014]
50. European Research Cluster on the Internet of Things: IoT Governance, Privacy and Security Issues. URL: http://www.internet-of-things-research.eu/pdf/IERC_Position_Paper_IoT_Governance_Privacy_Security_Final.pdf (2016). [Online; accessed Jan-'16]
51. Fehrenbacher, D.D.: Design of Incentive Systems. Contributions to Management Science. Springer (2013). DOI 10.1007/978-3-642-33599-0
52. Fipa, A.: Fipa acl message structure specification. Foundation for Intelligent Physical Agents, http://www.fipa.org/specs/fipa00061/SC00061G.html (2002)
53. Flinn, J., Sinnamohideen, S., Tolia, N., Satyanarayanan, M.: Data staging on untrusted surrogates. In: FAST, vol. 3, pp. 15–28. Citeseer (2003)
54. forgerock.com: Forge Rock. URL: https://www.forgerock.com/. [Online; accessed June-2014]
55. Frackowiak, G., Ganzha, M., Paprzycki, M., Szymczak, M., Han, Y., Park, M.: Adaptability in an Agent-Based Virtual Organization – Towards Implementation. In: J. Cordeiro, S. Hammoudi, J. Filipe, W. Aalst, J. Mylopoulos, N.M. Sadeh, M.J. Shaw, C. Szyperski (eds.) Web Information Systems and Technologies, vol. 18, pp. 27–39. Springer Berlin Heidelberg (2009). DOI http://dx.doi.org/10.1007/978-3-642-01344-7_3. URL http://www.springerlink.com/index/x6m368385hm12t36.pdf
56. Frank, B., Shelby, Z., Hartke, K., Bormann, C.: Constrained application protocol (coap). IETF draft, Jul (2011)
57. Franklin, M.J., Kossmann, D., Kraska, T., Ramesh, S., Xin, R.: Crowddb: Answering queries with crowdsourcing. In: Proc. 2011 ACM SIGMOD Intl. Conf. on Management of Data, SIGMOD '11, pp. 61–72. ACM (2011). DOI 10.1145/1989323.1989331

58. Frey, B., Jegen, R.: Motivation crowding theory. Journal of Economic surveys **15**(5), 589–611 (2001). URL http://onlinelibrary.wiley.com/doi/10.1111/1467-6419.00150/abstract
59. Galdon-Clavell, G.: (not so) smart cities?: The drivers, impact and risks of surveillance-enabled smart environments. Science and Public Policy **40**(6), 717–723 (2013)
60. Garavan, T.N., Carbery, R.: Collective Learning, pp. 646–649. Springer US, Boston, MA (2012). DOI 10.1007/978-1-4419-1428-6_136
61. Gilbert, N., Troitzsch, K.: Simulation for the social scientist. Open University Press, McGraw-Hill Education (2005)
62. Giunchiglia, F., Dutta, B., Maltese, V.: From knowledge organization to knowledge representation. In: ISKO UK Conference (2013)
63. Giunchiglia, F., Hume, A.: A distributed entity directory. In: The Semantic Web: ESWC 2013 Satellite Events, *LNCS*, vol. 7955, pp. 291–292. Springer Berlin Heidelberg (2013)
64. Goldstein, D.G., McAfee, R.P., Suri, S.: The wisdom of smaller, smarter crowds. In: Proc. ACM Conf. on Economics and Computation, EC '14, pp. 471–488. ACM (2014). DOI 10.1145/2600057.2602886
65. Guinard, D., Trifa, V., Karnouskos, S., Spiess, P., Savio, D.: Interacting with the soa-based internet of things: Discovery, query, selection, and on-demand provisioning of web services. Services Computing, IEEE Transactions on **3**(3), 223–235 (2010)
66. Gunkel, M.: Country-Compatible Incentive Design. DUV, Wiesbaden (2006). DOI 10.1007/978-3-8350-9214-3
67. Hardy, G.: Using IT governance and COBIT to deliver value with IT and respond to legal, regulatory and compliance challenges. Information Security technical report **11**(1), 55–61 (2006)
68. Hartswood, M., Jirotka, M., Chenu-Abente, R., Hume, A., Giunchiglia, F., Martucci, L.A., Fischer-Hübner, S.: Privacy for peer profiling in collective adaptive systems. In: Privacy and Identity Management for the Future Internet in the Age of Globalisation. Springer (2015)
69. Hassan, M.M., Song, B., Huh, E.N.: A framework of sensor-cloud integration opportunities and challenges. In: ICUIMC (2009)
70. Heckman, J., Smith, J.: What do bureaucrats do? The effects of performance standards and bureaucratic preferences on acceptance into the JTPA program. Advances in the Study of Entrepreneurship Innovation and Economic Growth **7**, 191–217 (1996)
71. Heneman, R.L., Dixon, K.E., Gresham, M.T.: Team pay for novice, intermediate, and advanced teams. Advances in Interdisciplinary Studies of Work Teams (7), 141–160 (2000). DOI http://dx.doi.org/10.1016/S1572-0977(00)07009-6
72. Hirth, M., Hoßfeld, T., Tran-Gia, P.: Analyzing costs and accuracy of validation mechanisms for crowdsourcing platforms. Math. and Comp. Modelling **57**(11–12), 2918 – 2932 (2013). DOI http://dx.doi.org/10.1016/j.mcm.2012.01.006
73. Hochschild, J.L.: If democracies need informed voters, how can they thrive while expanding enfranchisement? Election Law Journal: Rules, Politics, and Policy **9**(2), 111–123 (2010). URL http://www.liebertonline.com/doi/pdfplus/10.1089/elj.2009.0055
74. Hoisl, B., Aigner, W.: Social Rewarding in Wiki Systems – Motivating the Community. In: Proceedings of HCI International - 12th International Conference on Human-Computer Interaction (HCII 2007), pp. 362–371. Springer (2007). DOI http://dx.doi.org/10.1007/978-3-540-73257-0_40. URL http://www.springerlink.com/index/F864231372413836.pdf
75. Hollands, R.G.: Will the real smart city please stand up? intelligent, progressive or entrepreneurial? City **12**(3), 303–320 (2008)
76. Holmstrom, B., Milgrom, P.: Multitask principal-agent analyses: Incentive contracts, asset ownership, and job design. JL Econ. & Org. **7**(January 1991), 24 (1991). URL http://heinonlinebackup.com/hol-cgi-bin/get_pdf.cgi?handle=hein.journals/jleo7&section=31
77. Huang, S.W., Fu, W.T.: Don't hide in the crowd!: Increasing social transparency between peer workers improves crowdsourcing outcomes. In: Proc. SIGCHI Conf. on Human Factors in Comp. Systems, CHI '13, pp. 621–630. ACM (2013). DOI 10.1145/2470654.2470743

78. Huang, T.K., Ribeiro, B., Madhyastha, H.V., Faloutsos, M.: The socio-monetary incentives of online social network malware campaigns. In: Proc. ACM Conf. on Online Social Networks, COSN '14, pp. 259–270. ACM (2014). DOI 10.1145/2660460.2660478

79. IBM : SOA governance—IBM's approach. URL: `ftp://ftp.software.ibm.com/software/soa/pdf/SOA_Gov_Process_Overview.pdf`. [Online; accessed March-2017]

80. International Electrotechnical Commission: Orchestrating infrastructure for sustainable Smart Cities. URL: `http://www.iec.ch/whitepaper/smartcities/` (2016). [Online; accessed Jun-'16]

81. Irani, L.C., Silberman, M.S.: Turkopticon: Interrupting worker invisibility in amazon mechanical turk. In: Proceedings of the SIGCHI Conference on Human Factors in Computing Systems, CHI '13, pp. 611–620. ACM, New York, NY, USA (2013). DOI 10.1145/2470654.2470742

82. Jakumeit, E., Buchwald, S., Kroll, M.: GrGen. NET. Intl. J. on Software Tools for Technology Transfer **12**(3), 263–271 (2010)

83. Kaune, S.: Performance and Availability in Peer-to-Peer Content Distribution Systems: A Case for a Multilateral Incentive Approach. Phd thesis, TU Darmstadt (2011). URL `http://tuprints.ulb.tu-darmstadt.de/2492/1/Phd-Thesis-Kaune___2011.pdf`

84. Keller, A., Badonnel, R.: Automating the provisioning of application services with the bpel4ws workflow language. In: Utility Computing, pp. 15–27. Springer (2004)

85. Kerrin, M., Oliver, N.: Collective and individual improvement activities: the role of reward systems. Personnel Review **31**(3), 320–337 (2002). DOI 10.1108/00483480210422732. URL `http://www.emeraldinsight.com/10.1108/00483480210422732`

86. Kim, H., Feamster, N.: Improving network management with software defined networking. Communications Magazine, IEEE **51**(2), 114–119 (2013)

87. King, J., Bose, R., Yang, H.I., Pickles, S., Helal, A.: Atlas: A service-oriented sensor platform: Hardware and middleware to enable programmable pervasive spaces. In: LCN, pp. 630–638 (2006)

88. Kirkpatrick, K.: Software-defined networking. Communications of the ACM **56**(9), 16–19 (2013)

89. Kittur, A., Nickerson, J.V., Bernstein, M., Gerber, E., Shaw, A., Zimmerman, J., Lease, M., Horton, J.: The future of crowd work. In: Proc. of the 2013 Conf. on Computer supported cooperative work, CSCW '13, pp. 1301–1318. ACM (2013). DOI 10.1145/2441776.2441923

90. Kolbe, M., Boos, M.: Facilitating group decision-making: Facilitator's subjective theories on group coordination. Forum: Qualitative Social Research **10**(1) (2009). URL `http://nbn-resolving.de/urn:nbn:de:0114-fqs0901287`

91. Koldehofe, B., Dürr, F., Tariq, M.A., Rothermel, K.: The power of software-defined networking: line-rate content-based routing using openflow. In: MW4NG'12 (2012)

92. Kovatsch, M., Lanter, M., Duquennoy, S.: Actinium: A restful runtime container for scriptable internet of things applications. In: Internet of Things, pp. 135–142 (2012)

93. Kulkarni, A.: The Complexity of Crowdsourcing : Theoretical Problems in Human Computation (2011)

94. Kumar, K., Lu, Y.H.: Cloud computing for mobile users: Can offloading computation save energy? Computer **43**(4), 51–56 (2010)

95. Laffont, J.J., Martimort, D.: The Theory of Incentives. Princeton University Press, New Jersey (2002)

96. Lantz, B., Heller, B., McKeown, N.: A network in a laptop: rapid prototyping for software-defined networks. In: SIGCOMM Workshop on Hot Topics in Networks. ACM (2010)

97. Lazear, E.: Performance Pay and Productivity. American Economic Review **90**(5), 1346–1361 (2000)

98. Lazear, E.: Personnel economics: The economist's view of human resources. Journal of Economic Perspectives **21**(4), 91–114 (2007). URL `http://www.nber.org/papers/w13653`

99. Lee, K.: Augmented reality in education and training. TechTrends **56**(2), 13–21 (2012). DOI 10.1007/s11528-012-0559-3. URL `http://dx.doi.org/10.1007/s11528-012-0559-3`

100. Lewis, G., Echeverría, S., Simanta, S., Bradshaw, B., Root, J.: Tactical cloudlets: Moving cloud computing to the edge. In: Military Communications Conference (MILCOM), 2014 IEEE, pp. 1440–1446. IEEE (2014)

101. Leymann, F., Roller, D.: Production workflow: concepts and techniques (2000)
102. Little, G.: Exploring iterative and parallel human computation processes. In: Ext. Abstracts on Human Factors in Comp. Sys., CHI EA '10, pp. 4309–4314. ACM (2010). DOI 10.1145/1753846.1754145
103. Macal, C.M., North, M.J.: Agent-based modeling and simulation. Proceedings of the 2009 Winter Simulation Conference (WSC) pp. 86–98 (2009). DOI 10.1109/WSC.2009.5429318. URL http://ieeexplore.ieee.org/lpdocs/epic03/wrapper.htm?arnumber=5429318
104. Madden, S.R., Franklin, M.J., Hellerstein, J.M., Hong, W.: TinyDB: an acquisitional query processing system for sensor networks. ACM Transactions on database systems (TODS) **30**(1), 122–173 (2005)
105. Manin, B.: The Principles of Representative Government. Cambridge University Press (1997)
106. Mao, A., Kamar, E., Chen, Y., Horvitz, E., Schwamb, M.E., Lintott, C.J., Arfon M., S.: Volunteering Versus Work for Pay: Incentives and Tradeoffs in Crowdsourcing. In: Proc. First AAAI Conf. on Human Computation and Crowdsourcing, pp. 94–102. AAAI, Palm Springs, CA, USA (2013)
107. Martin, D., Hanraghan, B.V., O'Neill, J., Gupta, N.: Being a turker. In: Proceedings of the 17th ACM Conference on Computer Supported Cooperative Work and Social Computing, CSCW '14, pp. 224–235. ACM, New York, NY, USA (2014). DOI 10.1145/2531602.2531663
108. Martin Fowler: Microservices - a definition of this new architectural term. URL: http://martinfowler.com/articles/microservices.html. [Online; accessed Jan.-2016]
109. Mason, W., Watts, D.J.: Financial incentives and the "performance of crowds". In: Proc. ACM SIGKDD Workshop on Human Computation, HCOMP '09, pp. 77–85. ACM (2009). DOI 10.1145/1600150.1600175
110. Mendonca, M., Cowan, D.: Decision-making coordination and efficient reasoning techniques for feature-based configuration. Sci. Comput. Program. **75**(5), 311–332 (2010). DOI 10.1016/j.scico.2009.12.004. URL http://dx.doi.org/10.1016/j.scico.2009.12.004
111. Milinski, M., Semmann, D., Krambeck, H.J.: Reputation helps solve the 'tragedy of the commons'. Nature **415**(6870), 424–426 (2002). DOI 10.1038/415424a. URL http://dx.doi.org/10.1038/415424a
112. Minder, P., Bernstein, A.: Crowdlang: A programming language for the systematic exploration of human computation systems. In: K. Aberer, A. Flache, W. Jager, L. Liu, J. Tang, C. Guéret (eds.) Social Informatics, *LNCS*, vol. 7710, pp. 124–137. Springer (2012). DOI 10.1007/978-3-642-35386-4_10
113. Mohagheghi, P., Haugen, Ø.: Evaluating domain-specific modelling solutions. In: J. Trujillo, G. Dobbie, H. Kangassalo, S. Hartmann, M. Kirchberg, M. Rossi, I. Reinhartz-Berger, E. Zimányi, F. Frasincar (eds.) Advances in Conceptual Modeling – Applications and Challenges, *LNCS*, vol. 6413, pp. 212–221. Springer (2010). DOI 10.1007/978-3-642-16385-2_27
114. Mottola, L., Pathak, A., Bakshi, A., Prasanna, V.K., Picco, G.P.: Enabling scope-based interactions in sensor network macroprogramming. In: MASS 2007, pp. 1–9 (2007)
115. Mumby, D.K.: Theorizing resistance in organization studies: A dialectical approach. Management Communication Quarterly **19**(1), 19–44 (2005). DOI 10.1177/0893318905276558
116. Murray-Rust, D., Scekic, O., Papapanagiotou, P., Truong, H.L., Robertson, D., Dustdar, S.: A collaboration model for community-based software development with social machines. EAI Endorsed Trans. Collaborative Computing **1**(5), e6 (2015)
117. Nakamoto, S.: Bitcoin: A peer-to-peer electronic cash system (2008)
118. Nastic, S., Copil, G., Truong, H.L., Dustdar, S.: Governing Elastic IoT Cloud Systems under Uncertainty. In: The 7th International Conference on Cloud Computing Technology and Science (CloudCom 2015) (2015)
119. Nastic, S., Inziger, C., Truong, H.L., Dustdar, S.: GovOps: The Missing Link for Governance in Software-defined IoT Cloud Systems. In: WESOA14 (2014)
120. Nastic, S., Sehic, S., Le, D.H., Truong, H.L., Dustdar, S.: Provisioning Software-defined IoT Cloud Systems. In: FiCloud'14
121. Nastic, S., Sehic, S., Voegler, M., Truong, H.L., Dustdar, S.: PatRICIA - A novel programing model for IoT applications on cloud platforms. In: SOCA (2013)

122. Nastic, S., Truong, H.L., Dustdar, S.: Sdg-pro: a programming framework for software-defined iot cloud gateways. Journal of Internet Services and Applications **6**(1), 1–17 (2015)
123. Nastic, S., Truong, H.L., Dustdar, S.: A Middleware Infrastructure for Utility-based Provisioning of IoT Cloud Systems. In: The First IEEE/ACM Symposium on Edge Computing (2016)
124. Nastic, S., Truong, H.L., Dustdar, S.: Data and Control Points: A Programming Model for Resource-constrained IoT Cloud Edge Devices. In: IEEE International Conference on Systems, Man, and Cybernetics (SMC 2017) (2017). (To Appear.)
125. Nastic, S., Voegler, M., Inziger, C., Truong, H.L., Dustdar, S.: rtGovOps: A Runtime Framework for Governance in Large-scale Software-defined IoT Cloud Systems. In: Mobile Cloud 2015 (2015)
126. Niemann, M., Miede, A., Johannsen, W., Repp, N., Steinmetz, R.: Structuring SOA governance. International Journal of IT/Business Alignment and Governance **1**(1), 58–75 (2010)
127. OASIS: Device Profile for Web Services (DPWS) Specification. http://docs.oasis-open.org/ws-dd/ns/dpws/2009/01. [Online; accessed Jul-'13]
128. OASIS: MQ Telemetry Transport Specification. http://docs.oasis-open.org/mqtt/mqtt/v3.1.1/mqtt-v3.1.1.html. [Online; accessed Mar-'16]
129. Open Stack Orchestration: Heat Project. URL https://wiki.openstack.org/wiki/Heat. [Online; accessed Feb.-2015]
130. OpenStack.org: OpenStack – Open source software for creating private and public clouds. http://www.openstack.org/. [Online; accessed Mar-'14]
131. OpenTOSCA: OpenTOSCA. URL http://www.iaas.uni-stuttgart.de/OpenTOSCA/. [Online; accessed Feb.-2015]
132. OpsCode: Chef. URL http://opscode.com/chef. [Online; accessed Feb.-2015]
133. Pande, R., Banerjee, A.V., Kumar, S., Su, F.: Do informed voters make better choices? experimental evidence from urban india (2011). URL http://epod.cid.harvard.edu/publications/do-informed-voters-make-better-choices-experimentalevidence-urban-india
134. Pearsall, M.J., Christian, M.S., Ellis, A.P.J.: Motivating interdependent teams: Individual rewards, shared rewards, or something in between? Journal of Applied Psychology **95**(1), 183–191 (2010)
135. Peled, N., Gal, Y.K., Kraus, S.: A study of computational and human strategies in revelation games. Autonomous Agents and Multi-Agent Systems **29**(1), 73–97 (2015). DOI 10.1007/s10458-014-9253-5
136. Prendergast, C.: The provision of incentives in firms. J. of economic literature **37**(1), 7–63 (1999)
137. Price, D.D.S.: A general theory of bibliometric and other cumulative advantage processes. Journal of the American Society for Information Science **27**(5), 292–306 (1976). DOI 10.1002/asi.4630270505. URL http://doi.wiley.com/10.1002/asi.4630270505
138. Procaccianti, G., Lago, P., Lewis, G.A.: A catalogue of green architectural tactics for the cloud. In: Maintenance and Evolution of Service-Oriented and Cloud-Based Systems (MESOCA), 2014 IEEE 8th International Symposium on the, pp. 29–36. IEEE (2014)
139. Puppet Labs: Puppet. URL https://puppet.com/. [Online; accessed Feb.-2017]
140. Rao, H., Huang, S., Fu, W.: What will others choose? how a majority vote reward scheme can improve human computation in a spatial location identification task. In: B. Hartman, E. Horvitz (eds.) Proc. of the First AAAI Conf. on Human Computation and Crowdsourcing, HCOMP 2013, November 7-9, 2013, Palm Springs, CA, USA. AAAI (2013)
141. Reid, F., Harrigan, M.: An analysis of anonymity in the bitcoin system. In: Security and privacy in social networks, pp. 197–223. Springer (2013)
142. Resnick, P., Kuwabara, K., Zeckhauser, R., Friedman, E.: Reputation systems: Facilitating trust in Internet interactions. Communications of the ACM **43**(12), 45–48 (2000)
143. Retelny, D., Robaszkiewicz, S., To, A., Lasecki, W.S., Patel, J., Rahmati, N., Doshi, T., Valentine, M., Bernstein, M.S.: Expert crowdsourcing with flash teams. In: Proceedings of the 27th Annual ACM Symposium on User Interface Software and Technology, UIST '14, pp. 75–85. ACM, New York, NY, USA (2014). DOI 10.1145/2642918.2647409

144. Riveni, M., Truong, H.L., Dustdar, S.: On the elasticity of social compute units. In: M. Jarke, J. Mylopoulos, C. Quix, C. Rolland, Y. Manolopoulos, H. Mouratidis, J. Horkoff (eds.) Advanced Information Systems Engineering, *LNCS*, vol. 8484, pp. 364–378. Springer (2014). DOI 10.1007/978-3-319-07881-6_25

145. Rockenbach, B., Milinski, M.: The efficient interaction of indirect reciprocity and costly punishment. Nature **444**(7120), 718–723. DOI 10.1038/nature05229. URL http://dx.doi.org/10.1038/nature05229

146. Ron, D., Shamir, A.: Quantitative analysis of the full bitcoin transaction graph. In: International Conference on Financial Cryptography and Data Security, pp. 6–24. Springer (2013)

147. Rovatsos, M.: Multiagent systems for social computation. In: Proceedings of the 2014 International Conference on Autonomous Agents and Multi-agent Systems, AAMAS '14, pp. 1165–1168. International Foundation for Autonomous Agents and Multiagent Systems (2014)

148. Rovatsos, M., Diochnos, D.I., Craciun, M.: Agent Protocols for Social Computation. In: Second International Workshop on Multiagent Foundations of Social Computing (2015)

149. Rudin, C., Waltz, D., Anderson, R.N., Boulanger, A., Salleb-Aouissi, A., Chow, M., Dutta, H., Gross, P.N., Huang, B., Ierome, S., et al.: Machine learning for the new york city power grid. IEEE transactions on pattern analysis and machine intelligence **34**(2), 328–345 (2012)

150. Ryan, R.M., Deci, E.L.: Intrinsic and extrinsic motivations: Classic definitions and new directions. Contemporary Educational Psychology **25**(1), 54 – 67 (2000). DOI http://dx.doi.org/10.1006/ceps.1999.1020. URL http://www.sciencedirect.com/science/article/pii/S0361476X99910202

151. Sandrino-Arndt, B.: People, portfolios and processes: The 3p model of it governance. Information Systems Control Journal **2**, 1–5 (2008)

152. Sato, K., Hashimoto, R., Yoshino, M., Shinkuma, R., Takahashi, T.: Incentive Mechanism Considering Variety of User Cost in P2P Content Sharing. In: Global Telecommunications Conference, 2008. IEEE GLOBECOM 2008. IEEE, pp. 1–5. IEEE (2008). URL http://ieeexplore.ieee.org/xpls/abs_all.jsp?arnumber=4698201

153. Satyanarayanan, M., Bahl, P., Caceres, R., Davies, N.: The case for vm-based cloudlets in mobile computing. Pervasive Computing **8**(4), 14–23 (2009)

154. Scekic, O., Miorandi, D., Schiavinotto, T., Diochnos, D., Hume, A., Chenu-Abente, R., Truong, H.L., Rovatsos, M., Carreras, I., Dustdar, S., F., G.: Smartsociety – a platform for collaborative people-machine computation. In: Proc. of the 8th IEEE International Conference on Service Oriented Computing and Applications (SOCA'15). Rome, Italy (2015)

155. Scekic, O., Truong, H.L., Dustdar, S.: Incentives and rewarding in social computing. Comm. of the ACM **56**(6), 72 (2013)

156. Scekic, O., Truong, H.L., Dustdar, S.: Programming incentives in information systems. In: C. Salinesi, M.C. Norrie, s. Pastor (eds.) Advanced Information Systems Engineering, *LNCS*, vol. 7908, pp. 688–703. Springer (2013). DOI 10.1007/978-3-642-38709-8_44

157. Scekic, O., Truong, H.L., Dustdar, S.: Supporting multilevel incentive mechanisms in crowdsourcing systems: an artifact-centric view. In: Cloud-based Software Crowdsourcing, pp. 95–114. Springer (2015)

158. Scekic, O., Truong, H.L., Dustdar, S.: Pringl – a domain-specific language for incentive management in crowdsourcing. Computer Networks (9 July 2015). DOI http://dx.doi.org/10.1016/j.comnet.2015.05.019

159. Schaffers, H., Sällström, A., Pallot, M., Hernández-Muñoz, J.M., Santoro, R., Trousse, B.: Integrating living labs with future internet experimental platforms for co-creating services within smart cities. In: Concurrent Enterprising (ICE), 2011 17th International Conference on, pp. 1–11. IEEE (2011)

160. Seffah, A., Donyaee, M., Kline, R.B., Padda, H.K.: Usability measurement and metrics: A consolidated model. Software Quality Control **14**(2), 159–178 (2006). DOI 10.1007/s11219-006-7600-8

161. Segal, A., Simpson, R.J., Gal, Y., Homsy, V., Hartswood, M., Page, K.R., Jirotka, M.: Improving productivity in citizen science through controlled intervention. In: WWW, p. forthcoming (2015)

162. Semmann, D., Krambeck, H.J., Milinski, M.: Strategic investment in reputation. Behavioral Ecology and Sociobiology **56**(3), 248–252 (2004). DOI 10.1007/s00265-004-0782-9. URL `http://dx.doi.org/10.1007/s00265-004-0782-9`

163. Sengupta, B., Jain, A., Bhattacharya, K., Truong, H.L., Dustdar, S.: Who do you call? problem resolution through social compute units. In: Proc. 10th Intl. Conf. on Service-Oriented Comp., ICSOC'12, pp. 48–62. Springer-Verlag (2012)

164. Silberman, M.S., Nathan, L., Knowles, B., Bendor, R., Clear, A., Håkansson, M., Dillahunt, T., Mankoff, J.: Next steps for sustainable hci. interactions **21**(5), 66–69 (2014). DOI 10.1145/2651820. URL `http://doi.acm.org/10.1145/2651820`

165. Skopik, F., Schall, D., Dustdar, S.: Modeling and mining of dynamic trust in complex service-oriented systems. Information Systems **35**(7), 735–757 (2010). DOI 10.1016/j.is.2010.03.001. URL `http://linkinghub.elsevier.com/retrieve/pii/S0306437910000153`

166. SOA Software: Integrated SOA governance. URL: `http://www.soa.com/solutions/integrated_soa_governance`. [Online; accessed June-2014]

167. Soldatos, J., Serrano, M., Hauswirth, M.: Convergence of utility computing with the internet-of-things. In: IMIS, pp. 874–879 (2012)

168. Solomon, L.: Rethinking Our Centralized Monetary System: The Case for a System of Local Currencies. Praeger (1996). URL `https://books.google.at/books?id=8VRKVwP4JMsC`

169. Stuedi, P., Mohomed, I., Terry, D.: Wherestore: Location-based data storage for mobile devices interacting with the cloud. In: MCS (2010)

170. Thereska, E., Ballani, H., O'Shea, G., Karagiannis, T., Rowstron, A., Talpey, T., Black, R., Zhu, T.: IoTFlow: A software-defined storage architecture. In: SOSP, pp. 182–196. ACM (2013)

171. thingworx.com: ThingWorx. URL: `http://thingworx.com`. [Online; accessed Jan.-2015]

172. Tokarchuk, O., Cuel, R., Zamarian, M.: Analyzing crowd labor and designing incentives for humans in the loop. Internet Computing, IEEE **16**(5), 45–51 (2012). DOI 10.1109/MIC.2012.66

173. Tran, A., Zeckhauser, R.: Rank as an Incentive: Evidence from a Field Experiment (2009). URL `http://econpapers.repec.org/RePEc:ecl:harjfk:rwp09-019`

174. Tranquillini, S., Daniel, F., Kucherbaev, P., Casati, F.: Modeling, enacting, and integrating custom crowdsourcing processes. ACM Trans. Web **9**(2), 7:1–7:43 (2015). DOI 10.1145/2746353

175. Tridium: Sedona Virtual Machine. URL: `http://www.sedonadev.org/`. [Online; accessed Jan.-2016]

176. US President's Council of Advisors on Science and Technology: Report to the President and Congress ensuring leadership in federally funded research and development in information technology. URL: `https://www.whitehouse.gov/sites/default/files/microsites/ostp/PCAST/nitrd_report_aug_2015.pdf` (2015). [Online; accessed Jun-'16]

177. Van Herpen, M., Cools, K., Van Praag, M.: Wage Structure and the Incentive Effects of Promotions. Kyklos **59**(3), 441–459 (2006). DOI 10.1111/j.1467-6435.2006.00341.x. URL `http://doi.wiley.com/10.1111/j.1467-6435.2006.00341.x`

178. Vassileva, J.: Motivating participation in social computing applications: a user modeling perspective. User Modeling and User-Adapted Interaction **22**(1-2), 177–201 (2012). DOI 10.1007/s11257-011-9109-5. URL `http://dx.doi.org/10.1007/s11257-011-9109-5`

179. Vlacheas, P., Giaffreda, R., Stavroulaki, V., Kelaidonis, D., Foteinos, V., Poulios, G., Demestichas, P., Somov, A., Biswas, A.R., Moessner, K.: Enabling smart cities through a cognitive management framework for the internet of things. IEEE communications magazine **51**(6), 102–111 (2013)

180. Walker, C.C., Dooley, K.J.: The Stability of Self-Organized Rule-Following Work Teams. Computational & Mathematical Organization Theory **5**(1), 5–30 (1999). DOI 10.1023/A:1009689326098. URL `http://dx.doi.org/10.1023/A:1009689326098`

181. Waveworks: Wave router. URL: `https://github.com/weaveworks/weave`. [Online; accessed Mar.-2015]

182. Weber, K., Otto, B., Österle, H.: One size does not fit all—a contingency approach to data governance. Journal of Data and Information Quality (JDIQ) **1**(1), 4 (2009)
183. Weber, R.H.: Internet of things–governance quo vadis? Computer Law & Security Review **29**(4), 341–347 (2013)
184. Wedekind, C., Milinski, M.: Cooperation Through Image Scoring in Humans. Science **288**(5467), 850–852 (2000). DOI 10.1126/science.288.5467.850. URL `http://dx.doi.org/10.1126/science.288.5467.850`
185. Wegner, R.: Multi-agent malicious behaviour detection. Phd thesis, University of Manitoba (2012). URL `http://hdl.handle.net/1993/9673`
186. Xively: Xively. URL `http://xively.com`. [Online; accessed Jan-'15]
187. Yi, S., Li, C., Li, Q.: A survey of fog computing: concepts, applications and issues. In: Proceedings of the 2015 Workshop on Mobile Big Data, pp. 37–42. ACM (2015)
188. Yogo, K., Shinkuma, R., Takahashi, T., Konishi, T., Itaya, S., Doi, S., Yamada, K.: Differentiated Incentive Rewarding for Social Networking Services. 2010 10th IEEE/IPSJ International Symposium on Applications and the Internet pp. 169–172 (2010). DOI 10.1109/SAINT.2010.65
189. Yuriyama, M., Kushida, T.: Sensor-cloud infrastructure-physical sensor management with virtualized sensors on cloud computing. In: Network-Based Information Systems (NBiS), 2010 13th International Conference on, pp. 1–8. IEEE (2010)
190. Zaslavsky, A., Perera, C., Georgakopoulos, D.: Sensing as a service and big data. arXiv preprint arXiv:1301.0159 (2013)
191. Zeppezauer, P.: Diploma thesis: Virtualizing communications for hybrid and diversity-aware collective adaptive systems (2014). Wien, Techn. Univ., Dipl.-Arb., 2015
192. Zeppezauer, P., Scekic, O., Truong, H.L., Diochnos, D.I., Rovatsos, M., Schiavinotto, T., Carreras, I., Miorandi, D.: Smartsociety consortium, deliverable 7.1 - virtualization techniques and prototypes. `http://www.smart-society-project.eu/publications/deliverables/D_7_1/` (2014)
193. Zeppezauer, P., Scekic, O., Truong, H.L., Dustdar, S.: Virtualizing communication for hybrid and diversity-aware collective adaptive systems. In: Proc. of 10th Intl. Workshop on Engineering Service-Oriented Applications, WESOA'14, pp. 56–67. Springer (2014)

Printed in the United States
By Bookmasters